Lamarck's Revenge

Selected Books by Peter Ward

Rare Earth: Why Complex Life Is Uncommon in the Universe
(with Donald Brownlee)

The Life and Death of Planet Earth (with Donald Brownlee)

Gorgon: Paleontology, Obsession, and the Greatest Catastrophe
in Earth's History

Life as We Do Not Know It: The NASA Search for
(and Synthesis of) Alien Life

Out of Thin Air: Dinosaurs, Birds, and Earth's Ancient Atmosphere

Under a Green Sky: Global Warming, the Mass Extinctions of the
Past, and What They Can Tell Us About Our Future

The Medea Hypothesis: Is Life on Earth Ultimately Self-Destructive?

The Flooded Earth: Our Future in a World Without Ice Caps

A New History of Life: The Radical New Discoveries About the
Origins and Evolution of Life on Earth (with Joe Kirschvink)

LAMARCK'S REVENGE

HOW EPIGENETICS IS REVOLUTIONIZING OUR UNDERSTANDING OF EVOLUTION'S PAST AND PRESENT

Peter Ward

BLOOMSBURY PUBLISHING

NEW YORK · LONDON · OXFORD · NEW DELHI · SYDNEY

BLOOMSBURY PUBLISHING
Bloomsbury Publishing Inc.
1385 Broadway, New York, NY 10018, USA

BLOOMSBURY, BLOOMSBURY PUBLISHING, and the Diana logo are trademarks
of Bloomsbury Publishing Plc

First published in the United States 2018

ISBN: HB: 978-1-63286-615-8; eBook: 978-1-63286-617-2

LIBRARY OF CONGRESS CATALOGING-IN-PUBLICATION DATA

Names: Ward, Peter D. (Peter Douglas), 1949– author.
Title: Lamarck's revenge : how epigenetics is revolutionizing our understanding
of evolution's past and present / Peter Ward.
Description: New York: Bloomsbury Publishing, Inc., [2018] |
Includes bibliographical references and index.
Identifiers: LCCN 2018003831 | ISBN 9781632866158 (hardcover) |
ISBN 9781632866172 (ebook)
Subjects: LCSH: Epigenetics—History—Popular works.
Classification: LCC QH450 . W37 2018 | DDC 572.8/65—dc23
LC record available at https://lccn.loc.gov/2018003831

2 4 6 8 10 9 7 5 3 1

Typeset by Westchester Publishing Services
Printed and bound in the U.S.A. by Berryville Graphics Inc., Berryville, Virginia

To find out more about our authors and books visit www.bloomsbury.com
and sign up for our newsletters.

Bloomsbury books may be purchased for business or promotional use.
For information on bulk purchases please contact Macmillan Corporate and Premium Sales
Department at specialmarkets@macmillan.com.

For Steve Altabef: with thanks.

*Also for Michael Skinner at Washington State University: thank you.
From all of us. Because you have a no-thanks job that has to be done,
protecting our children from the chemicals of the Industrial
Nightmare.*

Contents

The Jurassic Park of Nevada

THERE is a scabrous mountainside in western Nevada that probably looks no different now than it did a century ago, when dusty miners hacked at the chaotically bedded rocks making up this geologically tarnished landscape. These last-chance men were looking for signs of metals in the sedimentary rocks, ores that were certainly not present when these strata were deposited as a shallow, tropical seabed between 225 and 190 million years ago. But millions of years later (yet still millions of years before the present time), the massive folding and compression of all of the North American Cordillera heaved these deeply buried strata and their enclosed fossils upward from their miles-deep grave, and amid that tectonic violence, cracks and crevices were produced in these rocks that were sometimes invaded by metal-rich fluids, rising from far deeper in the earth. These metalliferous fluids eventually turned to rock as well, but this time rock filled with gold, lead, and, most abundantly, silver. The result was the great discovery of the Comstock Lode and the announcement in 1859 followed by Eldorado Canyon, Austin, Eureka, and Pioche Mines, all discovered in the 1860s. The riches found here drew men from all over the globe, and the long-suffering women who followed them.

The Nevada miners searched for the telltales of silver-laced ore, that most precious of the state's mineral treasures. The dappled pits and darker mine shafts perforating the landscape remain like random polka dots of black and white and attest to uncounted tons of rock removed one pickax blow (or dynamite explosion) at a time. But in spite of their fervor and toil, few of those miners found anything but misery, and by the early twentieth century the land was given back to nature. Yet now, a century after the Nevada silver rush, a new breed of miner has come, but with goals far different from a bonanza strike. The riches they seek are information and data from the nature of the fossil record of these rocks. One of the best sites for their search is also one of the deepest ravines cutting into this mountainous landscape. Long ago it was rather facetiously named New York Canyon.

There is little that invokes New York, the state or city. The only Great White Way comes from the white limestone that reflects the merciless, year-round Nevada sunshine. There is certainly no silver in these rocks—not in this particular canyon, anyway. But there is scientific gold instead, information that can help answer a long-running scientific mystery.

If Charles Darwin had known about the fossil record of New York Canyon, he would have hated it, because this fossil succession would have contradicted his theory that fossils should exist as an "insensibly graded series"[1] of shapes demonstrating the slow change from one species to another. In fact, Darwin went to his grave knowing that in most cases fossil successions actually showed that the switch from one fossil species to another was not gradual at all. One kind of fossil species was overlain by an entirely different one, already cut in whole cloth.

The modern searchers of these rocks in New York Canyon are specialists in the fossil record and have come to test Darwin's theory, as well as to better understand one of the most consequential of all geobiological events, for the sedimentary record of this canyon and the surrounding regions gives evidence of one of the five largest of Earth's past mass extinctions,[2] events that were short-term annihilations of most kinds of life on Earth. Some come to see if there is anything about this ancient mass extinction that might yield wisdom about the Sixth Extinction, the one going on now. There is no disputing the fact that a gigantic mass extinction occurred throughout the world about 200 million years ago. But what happened soon after is the core of the mystery. In the time known as the Early Jurassic, a world emerged from catastrophe, a biologically bereft place with few species and few individuals of any life-form, save microbes. Darwinian theory cannot explain the fossil data. New species jumped from the graves of the old. How could mass death be followed by such rapid renewal of life?

Deep in the canyons here, hundreds of individual layers are bare of any fossils at all. But further up the wall are some of the most spectacular fossils of all time: the coiled, chambered shells of ammonites, themselves descendants of, and looking like, the still-living pearly nautilus. Stratum by stratum, specimen by specimen, the beautiful ammonite fossils are collected, numbered, and then later scrutinized with the powerful twenty-first-century means of quantifying morphology and its change. Even with means of studying morphological change far more powerful than was available to Darwin, the appearance of these species still seems to have been *sudden*. The term amply describes the appearance of species diverse in shape, abundant in number, and decidedly deficient in any kind of fossilized ancestors. And it is not just here in Nevada that this apparently rapid flowering of entirely new species decorates the oldest rocks dated to the beginning of the Jurassic period.[3]

At any global site with earliest Jurassic marine strata, the message is the same: New species appear with what seems like too much rapidity to be explained by current theory. This is a scientific problem not only for Darwin but for modern evolutionists, because the succession of fossils in these limestone canyons is inexplicable by Darwin's great theory of evolution alone and thereby challenges one of the most robust of all scientific understandings.

The revolutionists attacking the scientific bedrock that Darwin built are "evolutionists" as well, but they come armed with a new set of theories, from a field known as epigenetics. Some call themselves "epigeneticists"; others invoke another name and call themselves neo-Lamarckians.

Darwin's theory has undergone many modifications over the past 150 years. The "modern synthesis,"[4] a name given to the current version of Darwinian evolutionary theory, added the twentieth-century discoveries from genetics, molecular biology, developmental biology, and paleontology, among others, into the current theory of evolution. In fact, it is not one single theory but can be thought of as a "scientific paradigm,"[5] which is a collection of well-accepted theories. Other major scientific paradigms include the theories regarding relativity, quantum mechanics, and continental drift, each being composed of multiple interlocking theories combined into a whole. Like the others, evolutionary theory is entrenched and accepted. But on occasion, even seemingly irrefutable scientific principles and paradigms do falter because of revolutionary new discoveries that cannot be explained by the old theories.

The fossils from New York Canyon might turn out to be one of many lines of research helping to convince a conservative scientific establishment that evolutionary theory built only on Darwin's foundation is incomplete. What is missing are important new concepts from the field of epigenetics.[6]

To date, most discoveries adding to the science of epigenetics have come from modern biology, with little input from paleontology: It is rare indeed to find fossilized DNA where the marks left by epigenetic change are preserved. Yet epigenetics has a great deal to add to the overall understanding of the history of life, beginning with the origin of the first living species itself, and then continuing as a significant driver of the diversification into the many millions of species aggregated into the major categories of life defined today.

There is also a relatively new contention that, just as the major events in the history of life have been affected by heretofore ignored or undiscovered epigenetic processes, so too has the social as well as biological history of our own species been greatly affected by epigenetic processes. Thus not just our *evolutionary* history; human *cultural* history has been affected by epigenetics as well.

The crux of epigenetic theory is that major environmental changes occurring *during the life of an individual* can cause heritable changes *to that organism* during its lifetime that can then be passed on to the next generations. As a result of substantial *environmental* change experienced by the organism, a possible consequence can be *physical* changes to the organism's DNA and chromosomes. The environmental changes might be caused by chemical or other physical changes (such as loss or gain of oxygen and changes in temperature or water acidity or alkalinity, among so many others); by biological events, such as the onset of disease; or by new predators, loss of food sources, or many other factors. Humans are animals. War, famine, disease, domestic violence, drugs, cigarettes, or new kinds of chemicals in our food, water, air, agricultural fields—all of these can be the kinds of major environmental factors that can trigger genomic change by the addition of tiny molecules adhering to our DNA or through changes to the scaffolding holding the shapes of our DNA in ways that cause genes to turn on or off. And these are genes that would not have done this otherwise. Sometimes these changes happen only in a way that affects the organism in question. But sometimes they get passed on to future generations.

An increasing number of laboratory tests and experiments support the evolutionary pathway of "heritable" epigenetics: an event causing an organism to undergo a chemical change to its genome, usually (as noted above) through the attachments to DNA of tiny methyl molecules, each but a few atoms long. Yet when these seemingly insignificant hitchhikers plaster themselves onto a DNA molecule, changes in gene action can occur. Life-affecting chemicals coded by the genes of the organism might stop being made. Or new kinds of molecules might start being synthesized within a cell, chemicals that were not present before. Darwinian theory posits that genes are fixed, that nothing an organism can do during its life can affect its evolutionary and genetic legacy. But an increasing number of experiments show that environmental change taking place during the life of an organism can change not only the recipient but its descendants as well, making these heritable, epigenetic events prime movers of evolutionary change. Furthermore, they can cause *rapid* evolutionary change—more rapid by far than the slow, gradual change that Darwin posited as being caused by infrequent, randomly produced mutations. The process is not a rival to evolutionary theory: The process of heritable epigenetics is an addition to evolutionary theory. As such, it provides profound explanations for interpreting the fossil record, but perhaps, also, for evolutionary changes that have been produced by seminal moments in human history.

The history of life is composed of long periods of mostly slow and gradual environmental change, or no change at all for one or more millennia, and often the conformity of environments is mimicked by many of the communities inhabiting these static environments, themselves undergoing little compositional change in the kinds and relative numbers of species. But then comes the temporally short but radical change in those seemingly "permanent" conditions, causing oceans to become poisonous, or broad inland branches of the world's oceans to recede from formerly vast but shallow seas. Or far more rapid events, such as paroxysms of volcanism rapidly heating the atmosphere or, faster yet, the environmental effects of asteroid or comet impact. In similar fashion, human history seems to show analogous patterns, much as human warfare has been described for soldiers: long periods of boredom punctuated by short intervals of chaos, death, and destruction. A newer view is that both life on Earth, as well as human civilizations, have responded to these environmental catastrophes with evolutionary change far faster than during the calm periods, positing that sudden environmental stress to populations also stimulates epigenetic change in humans. For human civilizations, it is not the sudden change in oxygen levels or temperature, for instance, or a new kind of parasite or predator or competitor; it is the analogous events of war, famine, disease, and perhaps even religion that shake us and evolutionarily stir the pot of our species' gene pool.

Looking Back

A STAPLE of cinema, even from its earliest incarnations, has been the portrayal of the future and humanity's place in it, either overtly or allegorically, and quite often that vision has been dystopian. For example, consider the polluted cityscapes portrayed in the 1982 film *Blade Runner*, where it is small mom-and-pop stores that produce organs and whole creatures synthetically, while larger corporations produce artificial humans, or "replicants." The twenty-first-century movie sequel continued visions of the environmental and technological future where a technological elite equipped with godlike powers of invention produces products that eventually turn on their inventors (just as they did in *Blade Runner*, in the many *Jurassic Park* movies, and most recently in the television series *Westworld*).

We are still a long way (if ever) from building Turing-tested AIs so "human" that neither they nor we can tell that they are artificial, or from bringing dinosaurs back to life from the long dead. Yet the distant future often has a tricky way of arriving sooner than is comfortable. In one sense a "far future" that was technologically impossible prior to this new century has indeed arrived. We are building the laboratories and instruments now bringing to life new kinds of organisms that evolution never produced, and using these tools to concoct a welter of genetically altered or grown-in-test-tube animals and plants. We are now fully capable of artificially producing humans with attributes making them more efficient killing machines than any brought to life by natural selection. Superorganisms. The means to build them comes from a theory first espoused almost a quarter of a millennium ago by the French naturalist Jean-Baptiste Lamarck, using a new term for that science: *heritable epigenetics.*

Beyond imagining what the future might hold, television and cinema have two prime motives: to make money for large corporations and to entertain the masses. Yet beyond all of our wishes to be entertained is a third role of big-budget screen entertainment: as a refuge from stress. There is a palpable sense of fear that the near future evokes, because never before has technology been so frightening to so many people. It is no longer simply the possibility of nuclear Armageddon that can keep children fearful in the night. Biology is now more threatening and at the same time more promising for our next generations. Designer soldiers can be faster, stronger, deadlier. Designer children can be smarter, healthier, more beautiful, more long-lived. Biology is the curse and the blessing, and as a main purveyor of our species' emotions, cinema knows this. Now cinema is economically dominated by humans that are "super." Some are good, some are evil. All are more powerful than we "ordinary" (i.e., produced by evolution) members of *Homo sapiens*. They are also subtly portrayed as what *we* need to become to survive this increasingly violent, crowded, toxic world. And watching them on-screen can keep the nightmares at bay, at least in two-hour shots.

We want to be entertained, which is often synonymous with escape, because outside of the multiplex or our various screens at home as well as at work, the world *is* getting scarier. Going outside is more dangerous. Staying home is safer. Our screened world is the safest place of all for many of us. The screened world, be it in the multiplex, the home TV, the iPad, or the cell phone, is also a place where our species evolves culturally—and, according to many scientific seers, probably biologically as well. The average American spends a minimum of ten hours on one kind of screen or another each day.[1] Now the same movies can be delivered to us at the touch of a button, and that touch can serve as a means of isolating ourselves from the human community. Where once there were suits of armor to defend ourselves, now we are armed with cell phones, and this transition may be rapidly evolving the human race.

Are our genes changing as fast as culture and technology? More important, does *anything* we experience during our lifetimes have any effect on our own genomes, our inherited genes, the information locked in our DNA that has been uniquely ours since birth? Based on Darwinian evolution, now called the "new synthesis," the answer is a reassuring and resounding "No!" It is an answer megaphoned by leading scientists who keep the flame of the Darwinian paradigm alive, and backed by the major scientific funding agencies. Yet epigenetics argues otherwise.

Darwin and his great theory have always seemed to give a grace note of safety: that our genes are impervious to change during our lifetimes. Biologically, this meant that no matter how badly you screw up via bad choices, such as the use of drugs, cigarettes, or alcohol, or exposure to toxins, violence, religion, or love, none of it will affect the genes you pass along to your children.

Thus according to current evolutionary theory, events in our lives, both the good (achieving happiness, religious contentment) and the bad (acquiring post-traumatic stress disorder from exposure to inhuman violence, or having been abused as a child, or growing up near a factory releasing polychlorinated biphenyl [PCB]-like poisonous chemicals into the nearby environment), *are meaningless to the children we might produce.*[2] Darwin gave us this solace: Nothing that happens in our lives can affect what we pass down to our children through heredity. The revolution that is epigenetics demonstrates that this is not true.

Charles Darwin espoused evolution as driven by natural selection. However, an earlier theory, proposed more than a half century before the first publication of Darwin's greatest work, came from a naturalist whose life and work were limned by the flames of the French Revolution.

Jean-Baptiste-Pierre-Antoine de Monet, Chevalier de Lamarck, had a different view about heredity and why animals changed through time. His scientific beliefs were that things that happen to us during our lives *can* change what we pass on to our next generation, and perhaps into even further generations. Darwin knew well what Lamarck theorized. Darwin believed that his own theories about evolution could not coexist with any aspect of what Lamarck postulated. We now know this is no longer the case.

Lamarck's Revenge looks anew at what are, perhaps, humanity's most basic questions: the "where," "when," and "why" of getting to the present-day biota on this planet. But the vehicle to do this is by asking specifically about the "how." What were the evolutionary mechanisms, the balance between Darwinian and neo-Lamarckian (aka heritable epigenetics), that produced not only our physical biology but some aspects of our heritable behavior as well?

Here are some possibilities. First, that the process known as epigenetics combined with periods of extraordinary environmental change has played a far greater role in what is called the "history of life" than is accepted by all but a small cadre of revolutionary biologists. This is perhaps most decisively shown through the epigenetic process of "lateral gene transfer,"[3] where on a given day, in a given minute, some organism is invaded by another and a product of that invasion is the incorporation of vast numbers of new genes, making the invaded creature something else again, neither the invader nor the invaded. This is known.

Second, new evidence[4] points to a probable role of epigenetics in producing rapid species transitions by mechanisms other than lateral gene transfer. Science has discovered that major evolutionary change of a species can happen a thousand times faster by epigenetics than by the process demanded by the Darwinian theory of single, random mutations along a creature's genome or DNA (or, in some cases, RNA). This is most likely to occur during and immediately after rare, major environmental perturbations (such as mass extinctions and their aftermath).

Many scientists believe we are in such a period again, and that humanity itself is surrounded by genomes undergoing "epi-mutations," the extremely rapid change of genomes by the major epigenetic processes, themselves triggered by environmental crises during random day-to-day existence.[5] It makes sense that we are not only surrounded by such change but that our own genes are equally malleable and now equally affected. In humans, such crises work through the effects of our mammalian stress systems, which are coupled to human gut biomes. It has been our response to cancer-causing environmental toxins and our responses to war, famine, disease, and strident religion; to the poisons we eat; the poisons we hear on partisan media; the poisons we bear through racism, sexism, and any form of abuse, from child to spousal to bullying in general. Stress hurts us. Stress also changes us epigenetically. We evolve from stress,[6] and we pass on new characteristics acquired during our lifetimes.

The many physical environments or habitats colonized by life are obviously not the same, with some being more rigorous than others. But in exploring many of the veritable libraries written about evolutionary theory, what seems to often be missing concerns the intersection of time and environment.

Yes, there are genuine paradises for Earth life, places like the rain forests and corals reefs so filled with the ingredients that support life that they are packed with species, and have been since the time of the first animals on Earth. At the opposite ends, in the most inhospitable places on land and in the greatest depths of the oceans, there are far fewer species. In similar fashion, some time periods have been more challenging to life than others, even in the most supportive of environments. There are good times and bad on Earth, and it is proposed here that that dichotomy has fueled a coupling of times when evolution has been mainly through Darwinian evolution and others when Lamarckian evolution has been dominant. Darwinian in good times, Lamarckian in bad, when *bad* can be defined as those times when our environments turn topsy-turvy, and do so quickly. When an asteroid hits the planet. When giant volcanic episodes create stagnant oceans.

When a parent becomes a sexual predator. When our industrial output warms the world. When there are six billion humans and counting.

The history of humanity also has witnessed fluctuations in "environmental" conditions through time. Such stress might be quantified in some respects—theoretically, in the average level of stress hormones of a human at a given moment in time. Environmental changes range from the coming and going of the repetitive ice age advances of the last 2.5 million years to the times of global disease and pestilence, or global hunger, or global war, or even heightened level of violence. Have these ebbs and flows caused variance in the rate of evolutionary change of our own species by triggering rapid epigenetic evolution, compared to the more tranquil periods, when change—if it took place at all—was the slower, more Darwinian kind? What if we could take a sample of global human stress in the same way that paleontologists take a sample of global organismal diversity (number of species) and disparity (number of body plans) at some interval of geologic time? In this experiment we would compare stress levels continent by continent, race by race, gender by gender, age by age. What is the level of stress molecules in descendants of enslaved people or survivors of the Holocaust or genocide? What is the level of stress in the rich compared to the poor? Which groups are evolving more quickly at least partially by heritable epigenetics? And most important: If stress in our modern world is causing human evolutionary change, what are we evolving into?

These are uncomfortable questions. But comfort is not something science cares about. Scientists have actually posed these questions,[7] and by sampling both human and animal bones from the last several millennia we can measure the level of epigenetic change for given times. Scientists in the emerging field of paleophysiology are searching the entire archeological record, sampling the bones of man and beast in search of answers. How much epigenetic change will be visible from the extracted DNA?

There is far more to evolution than simple morphological or physiological change. Behavior—violence, religion, sexism, love, tolerance, racism, intolerance—can be hereditary in at least having the *capability* to change genomes. All of these might be changing our species. The balance of hormones in each of us is affected by our exterior experiences; all that we experience during our lives can affect the generations we contribute to. Lamarck first suggested this. That it is not just whether we survive our environment, but what our environment does to us. Now we know that this is indeed the case. Our DNA is changed not by subtraction or addition of new code, but by adding on tiny molecules that act like on-off switches. Genes that once worked no longer do. Genes that were switched off by natural selection get switched back on.

CHAPTER I

From God to Science

THE earliest incarnation of the ideology that would become the theory of heritable epigenetics has its origin at the turn of the nineteenth century, when it was described by Jean-Baptiste Lamarck as what later biologists would call "the acquisition of an 'acquired characteristic.'"[1]

In the later 1800s, Lamarckists continued their namesake's quest to observe nature and to answer such questions as: Could a species actually change, or was it forever fixed in morphology and, perhaps more crucially, in behavior? Could the morphological makeup of the many kinds of life on Earth be the result not of the hand of a Creator but of environmental changes affecting that life?

Ultimately, Lamarck arrived at a three-step process[2] in what was to be the first really rational explanation for what we now call "organic evolution." First, an animal experienced a radical change of the environment around it. Second, the initial response to the environmental change was some new kind of behavior by that animal (or whole species). Third, the behavioral change was followed by morphological changes that were heritable in subsequent generations. This proposed process came to be named after its author. Today, a variant on what Lamarck proposed is sometimes called "neo-Lamarckism," but more often "epigenetics," or "heritable epigenetics."

By the 1860s, however, Lamarckism was discarded[3] in favor of an explanation first elucidated by Charles Darwin: that most evolutionary change is the product of natural selection. But now there has been a rebirth of the understanding that some kinds of evolutionary change took place in the past, are taking place in the present, and will take place in the future that are close in spirit if not scientific specifics to Lamarckism. (In Lamarck's and even Darwin's time, there was no field of genetics, or knowledge of DNA and RNA.)

Epigenetics is a subset of evolution. It is a process that causes some specific evolutionary changes. To some it is just a minor tweak of already understood processes and of little importance in the broader scheme not only of evolutionary change but of the past and even future history of life. But to others it is a process still poorly understood but potentially of far greater importance to mainstream evolutionary theory than strict Darwinists admit. To a few, its ongoing discovery is causing an unfolding scientific revolution. It has almost never been tied to the fossil record.

The rise and fall of Lamarck's theory of evolution is scientific history. But in originating and then promoting his novel ideas, Lamarck challenged the establishment of the best-known naturalists of his time, and for his ideas he was hounded to the point of loss of money, reputation, and then health. Yet, novel as his ideas were at the time, they did not arise out of a scientific vacuum. Like the ideas (and work) of scientists today, Lamarck's ideas were built upon previous foundations of thought.

To many, the word *science* conjures reams of organized information already known. But science is more of a template for action than the sum of its discoveries. Originally, it was philosophers, not the early scientists themselves (who commonly referred to themselves as "natural philosophers"), who provided the rules for those actions.

But how did humanity get to these competing theories of evolution at all? To build the scientific-discovery toolbox needed by the early evolutionists, we need to step back more than three centuries. The discoveries of science are now organized into disciplines, such as astronomy, biology, chemistry, and physics. Each of these has principles (now unquestioned fact), theories (probably true), and hypotheses (to be tested and discarded at the drop of a datum). Sometimes principles and theories cross numerous disciplines. They are like dictator-run governments: They exist until, or if ever, they are overthrown in a revolution.[4]

EVOLVING SCIENCE ITSELF

The highest level of scientific acceptance traditionally has been called a "principle," or sometimes a "law." Any branch of science has these as foundations: the principle of relativity in physics, of quantum mechanics in chemistry and physics; the principle of uniformitarianism in geology; and the principle of evolution in biology, among others. Yet there is also another way of looking at each of these disciplines, not as the sum of their laws and principles, but as active spheres of research driven under an aegis of understanding and striving to strengthen that accord, a pact silently accepted by the workers toiling in the given field. The great historian of science Thomas Kuhn called these "paradigms."[5]

The supposedly strongest and least changeable pillars of any scientific discipline, paradigms in general combine more than a single law or principle in such a way as to provide scientific unification under a single conceptual umbrella, but also to guide future research. Yet the parallel is of a feudal landholder in the Middle Ages, and many have noted that, while considered among the most modern of humanity's constructions, the current system of science is among the last to utilize the medieval system of apprentice and master. Graduate students are apprentices, spending five or six years earning a pittance while observing and learning their craft, at the same time providing work (data) for the benefit of the craftsman (the PhD graduate supervisor).

The analogy to the Middle Ages of modern universities goes beyond that (and beyond the obvious physical imitation of university buildings to the medieval cathedrals, whose monks and Catholic Church intellectuals were the only light of learning in those millennium-ago times).

The serfs (scientists in the modern analogy) toiled to provide more wealth for the feudal lord, getting a living and some level of protection for their work. But work it was, and the goal of the work, whether tilling fields or building higher and thicker walls of the castle, was to increase the strength of the lord of the manor. Work that in any way cast doubt on the validity of the system was violently extirpated. And so too with science. The lords of the manor of the major branches of science control the spigot of grant funds, the power of bestowing patronage through jobs, graduate students, and honors. There is real power at play, especially in this world of billions of dollars of scientific funding.

Paradigms can almost be thought of as living, selfish creatures, guiding the many scientific acolytes to ever further enhancement through experimentation, observation, modeling, and analysis in well-read reviews. Paradigms are only killed by regicidal acts, which Kuhn labeled "revolutions." Examples are many: the replacement of the Earth-centric model of the solar system by the Copernican system, the replacement of Newtonian physics with relativity, the replacement of the expanding Earth model with plate tectonics, among many others. When one system falls, there is a period of scientific instability. And intellectual violence, for much is always at stake—from the simple human emotion of pride to the simple human need of putting food on the family table. Vast sums of money run the global scientific enterprises and in relative terms always have since the rise of modern science in the late eighteenth century, whether those money sums came from crown, nobility, or national scientific, taxpayer-derived foundations. No paradigm dies without struggle, neither do its adherents give up without fighting.

Scientific revolutions rarely occur quickly, or from a single attack. Someone, somewhere, makes an observation that absolutely cannot be explained by the current paradigm. Generally, the new attempt to explain the otherwise inexplicable new observation is attacked as incorrect. But if the idea is confirmed as valid, a small flame quickly

spreads to a larger conflagration. Sometimes the grating new data
turn out to be from dishonesty (Piltdown man) or from honest but
brutally wrong mistakes (cold fusion). Sometimes the new data are
from neither of those, and cannot be explained under the current paradigm's umbrella. Amid the fights that ensue, the protectors with too
much to lose should the current paradigm fall will go through the same
stages of grief attached to one learning of a fatal illness. Eventually
there is acceptance.

Evolution is no different. Some believe that we are in the first phase
of a scientific revolution, one that threatens the current paradigm given
multiple names, but most commonly known as "Darwinian evolution."
The major principles or laws of evolution come from evidence culled
from the combined disciplines of genetics, biochemistry, and population ecology, among others. Epigenetics is one kind of evolution, if we
most simply define *evolution* as the change in a species of organisms
through time. Epigenetics is one way such changes occur.

The favored theory of Charles Darwin, and the "Darwinians" who
followed him, is that the major process of evolution is driven by natural
selection combined with genetic change by mutation.[6] Epigenetics
posits that a quite different set of circumstances can drive evolutionary
change, and that both Darwinian- as well as epigenetic-driven change
(or, to do him honor, Lamarckian-driven change) can proceed simultaneously. But the question still unanswered is simply knowing which
predominates. Or when. It is clear both exist. Increasingly, it appears
that the paradigm of Darwinian evolution cannot explain new findings
showing how environmental stresses during the life of an organism
can indeed produce biological consequences in the progeny of the individual being buffeted by environment.

Near the start of the second decade of this twenty-first century,
there was increasing strife between the establishment evolutionists and
those pushing for the taboo recognition of Lamarckian interpretations
of new data.[7] For not only the students taking courses about evolution

today but also those coming a century before, Lamarckian evolution has been derided. It was a theory of historical curiosity, but, at its core, the idea that any organism could pass on genetic information accumulated during its life was considered scientific fallacy.

The very antagonism remains surprising. As more and more data accrued, each increment of epigenetic understanding seemed profoundly unexplainable using Darwinian evolutionary theory alone.

The Kuhnian paradigm that we call Darwinian evolution, or sometimes biological evolution, is a fundamental scientific understanding. It transcends what we call "science" by seeping out of scientific journals, or even popular science books, then books and magazines and ultimately websites into human culture. It does this perhaps more so than any of these other time-tested descriptions of how the natural world is constructed. Not only all scientists (creationists, or those who espouse intelligent design, are not scientists!) but also most mainstream religions as well as educated laypeople intuitively understand and accept it. Modern movies such as those in the *Jurassic Park* franchise pack their punch because we understand that dinosaurs lived hundreds of millions, not hundreds of years ago. And like so many of the other great foundations of science, much of our understanding and acceptance that organisms *have* changed over time, that they are mutable, comes from discoveries of the past two centuries.

FROM NATURALISTS TO GEOLOGISTS
TO EVOLUTIONISTS

For more than a thousand years, philosophers[8] recognized that there was known regularity in the movement of stars and planets, the change of days, the workings of weather, and so many other eminently observable manifestations of nature. Many also were sure that there was more to explain the presence of so many different kinds of animals and plants than giving in (or giving up) to divine explanations. And unlike those who observed the stars and planets and pondered their movements, those who raised the animals and plants so necessary for survival through agriculture gave elegant testimony of organic change through time. Even our own aging was itself testimony to the restless and changing dictates of time on nature.

The triumph of evolutionary theory over divine explanations was immeasurably aided by the eighteenth- and nineteenth-century rise of the field we now call geology.[9] The observation of rocks and their organization, and especially the discovery of the fossils within them—lithic curiosities that ranged in shape and form from being nearly identical to still-living organisms to bizarre but still identifiable as once-living objects—was a great leap in evolutionary thought. The abundant trilobites, ammonites, graptolites, and crinoids, among so many other kinds of fossils—let alone the magnificent ichthyosaur skeletons showing combinations of traits seen in fish and lizards—all combined to fertilize the ground from which evolutionary theory would grow. It was plain that there were forms found in rocks different from the kinds of animals found at the time, and plain as well that rocks claimed a kind of antiquity that was intuitively far older than the ages claimed by the likes of James Ussher,[10] a seventeenth-century Anglican church elder of Northern Ireland, who dated the Earth as beginning in 4004 B.C.

The rock record gave us form, and changes in form gave us a sense of time passing, in the long ago. Study of how deeply antique the world was (and is) was undertaken mainly by curious naturalists, most of whom were wealthy landowners with time on their hands. From the dawn of the Renaissance onward, great clashes had been taking place between the Catholic Church and astronomers such as Galileo and especially Copernicus about Earth's (and, through analogy, mankind's) place in the cosmos. Those who studied fossils and deduced the great length of geologic time were pitted against those who followed the supposed information given by the Bible, in the case of the Christians, and by other ancient texts among Hindus and followers of Islam.

It was the birth of geology, mainly in Europe, that brought the religious leaders out of their somnolent hubris in believing that all knowledge was to be found in the Bible. That Archbishop Ussher found it necessary at all during his lifetime (1581 to 1656) to make his famous study and pronouncement on the age of the Earth was a direct response to naturalists such as Nicolaus Steno,[11] who in the same time period as Ussher would give science his bedrock principles of stratigraphy. He proclaimed that in piles of sedimentary rock, composed of layers laid down one at a time in temporal sequence, the lowermost were older than those above.

Steno also made the first really accurate pronouncements about fossils, using the presence of lithified sharks' teeth from Italy to show that fossils should be differentiated from gems and other kinds of rocks coming from the Earth, but even more important, showing the relationship between these sharks' teeth turned to stone and the teeth of sharks living in the seas off Italy, as well as elsewhere in the oceans.[12] The concept that living animals could die and be turned to stone and that that stone itself is older than any extant life seems trivial today, but it was of immense importance in building what would become the geologic timescale. Without time, there can be no concept of evolutionary change. These and other observations enraged the clergy and the Church.

Ussher was the first among either the nascent geologists or the clergy to make an actual estimate of the age of the Earth. By counting the generations of the Bible and adding them to modern history, he fixed the date of creation at October 23, 4004 B.C.[13] Later, John Lightfoot of Cambridge University in England proclaimed that the time of creation was 9:00 A.M. on October 26, 4004 B.C., which is about as precise as one can get.

This religious conclusion that the Earth and life were only about six thousand years old supported a prevalent theory known as the Great Chain of Being,[14] holding that God created a continuous series of life-forms infinite in number and variety, each one grading into the next, from simplest to most complex. In fact, this idea verges on many now-accepted concepts that we understand as biological evolution in the sense of the presence of gradations of complexity. But where it collided with the hard-won findings of the naturalists of the seventeenth and eighteenth centuries was that all of these many forms, including humans, were made by God, and made relatively recently. Furthermore, they were made in their present form and then never changed. The gauntlet was thrown down. There was no such thing as biological evolution, and in this climate the study of biology became stilted and confined to attempts to describe and classify animals and plants, with the end results of naming them but not trying to see the relationships between them, which we see now are clear evidence of evolutionary affinity, and thus of evolutionary change itself.

The seventeenth- and eighteenth-century naturalists were led in this time toward strict classification through the efforts of the greatest classifier of them all, the mid-eighteenth-century Swedish botanist Carolus Linnaeus (Carl von Linné), whose life work of 180 books gave us the formal names of so many animals and plants still accepted today.[15] He also made the enormous contribution of developing a construct to give each distinct organism two names, a genus and a species, and also creating a hierarchical system of taxonomic categories. (Taxonomy is the science of defining species into evolutionary lineages and categories.) A species belonged to a genus, usually with other species as well. Genera were placed in families, families in orders, and on through classes, phyla, and kingdoms.

Yet, for all of this, the books of Linnaeus are filled with precise descriptions of nature but say little else: He did little analysis or inter-pretation. Linnaeus believed that the enormous quantity and variety of life that he and his followers so meticulously described and classified was a product of an unchanging order of life created by the single God in heaven. All of this vast work was done as a testament to God. At least, that is how he rationalized his good work for most of his life. But while the naturalists bent on classification preserved their plants on beautiful bound pages, or mounted bones from animals and then described the dead, the many practical men and women staying alive through agriculture were making quite different kinds of observations.

It was quite clear to those in the world of animal husbandry that the Linnaean dictates of unchanging species was ludicrously wrong, based on the rapid changes they saw in animals as well as plants under cultivation. But an even more obvious example surrounded the Europeans: the various breeds of dogs that came into existence not by God but by humans. Too late, aging fast amid the nagging pains of old age, and knowing that he had little time to live, Linnaeus fixated on the well-known agricultural fact that plant hybrids, crosses between kinds of plants described by him as separate and thus unchanging species, were changing rapidly by induced cross-pollination. The results of many of these crosses were varieties that had not existed before. Linnaeus could not bring himself to communicate to other naturalists what was widely known by farmers: That species are not immutable. That even in a few generations one species could change into something radically and observably different. These plants had evolved.

If in his lifetime Linnaeus failed to admit what he must have known—that biological evolution was a reality—he did leave the conceptual framework that would grow into a model overtly based on evolutionary change: his hierarchal system of classification.[16] Following his death, other naturalists quickly embraced the methods, understanding that animals and plants placed in any of the taxonomic categories were biologically more similar to each other than to those in other, equivalent taxonomic levels. For example, all primates, including humans, are placed in a single order.

It is clear that all the species in this order share many similarities—far more similarities, in fact, than, say, with the animals making up the order Carnivora, which includes various kinds of cats, wolves, weasels, and others. Yet, how could those similarities have come about? Did God not actually create a Great Chain of Being in which species were morphologically fixed but rather a system in which life could be bunched in groups categorized by shared traits?

Linnaeus left a famous quote that reverberated through time to his disciples, and then to the famous naturalists struggling with the idea of "If not God, what?" Linnaeus noted, "*Natura non facit saltum*," which roughly translated means "Nature makes no leap." In this Linnaeus indicates that there should not be discontinuities between the varieties of life, and nearly a century later, Charles Darwin would explicitly quote this in arguing for the gradual nature of evolutionary change.

By the middle of the eighteenth century, the time was right for naturalists to accept what was shown in the immense amount of evidence amassed equally by the classifiers and the agriculturalists—that, in fact, species were not necessarily unchanging through time. That life was not "fixed" in form. Of the freethinkers who began to write and publish these heretical ideas (in the opinion of the still-powerful Catholic Church), none was more important than the aristocrat and brilliant French mathematician and naturalist Georges-Louis Leclerc, better known to us as the Comte de Buffon,[17] generally shortened to Buffon. It is no coincidence that Buffon was a botanist, as by that time the evidence of what we now recognize as rapid evolutionary change was most observable in the crossing and breeding of crop plants.

Buffon was close to the breakthroughs that our time almost universally credits to Charles Darwin. Buffon, and Darwin later, was convinced that living things do change through time. He believed that this was somehow a result of influences from the environment, or even chance. The latter idea would not be accepted until our time, with the work of Stephen Jay Gould and David Raup, paleontologists who provided convincing commentary on the role that random chance has played not only in many evolutionary events but in the history of life as a whole, perhaps best illustrated by the drastic change in Earth life caused by a rare chance impact of a large asteroid on Earth about 65 million years ago.

Buffon had many other convictions that were novel and daring in his time, and taken as fact in ours. He believed that the earth must be much older than 6,000 years. In 1774, in fact, he speculated that the earth must be at least 75,000 years old. He also suggested that humans and apes are related.[18]

Buffon anticipated Charles Darwin in many things, one of which was the innate understanding of the dangers of unfettered population growth in any "species" (although that concept was still hazy at this time). Buffon wrote the seed of a common tree could produce in as little as ten years a thousand new seeds (depending on the tree; Buffon used the European elm as his example). If all were allowed to grow, he speculated that our world would be covered with this particular tree, all coming from one seed, in no more than 150 years. He made similar thought experiments with fowl and fish as well. But for the latter he also noted that, in the case of such runaway population growth, there would be a horrific price to pay sooner than later from starvation and disease among the conquering population.

Though scientifically daring, Buffon was no fool. In the last years of King Louis XVI and the French monarchy, Buffon knew that many of his scientific conclusions would be not only controversial but also politically dangerous if they became widely known. He was a nobleman, and the aristocrats of the time were well aware of the anger among the peasantry from whom they took their wealth by virtual slavery through serfdom. Buffon was careful to hide his radical views in a limited-edition forty-four-volume natural history book series called *Histoire naturelle, Générale et particulière* (1749–1804). By doing this, he avoided broad public criticism.

In hindsight, Buffon made lasting contributions, and as we shall see, he was the springboard to two subsequent thinkers: One was his contemporary (Lamarck), while the other was Charles Darwin, who came to fame decades after both of these French thinkers were dead. Perhaps Buffon's most significant contribution was his insistence that natural phenomena must be explained by natural laws rather than theological doctrine. He was also an early and effective advocate of the Linnaean classification system, as well as a pioneer in asserting that species can change over generations.

Yet, as advanced as he was, there was one conclusion that was a bridge of discovery too far. Buffon publicly rejected the idea that species could evolve into other species, and, in so doing, he missed his chance to become what Charles Darwin did ascend to: the father figure of evolutionary theory. (In one respect it seems just as well to me. Instead of Darwinism, we would have "Buffonism." That is far too close to "Buffoonism." Evolution has enough public relations problems as it is!)

THE FIRST DARWIN

Another late-eighteenth-century closet evolutionist was Erasmus Darwin, the grandfather of Charles Darwin. Erasmus believed that evolution occurred in living things, including humans, but he could only speculate about what might be the cause. He wrote of his ideas about evolution in poems(!) and in a publication titled *Zoonomia; or, the Laws of Organic Life* (1794–96).[19] In the latter work, he also suggested that Earth and life on it must have been evolving for immense periods of time:

> Would it be too bold to imagine, that in the great length of time, since the earth began to exist, perhaps millions of ages before the commencement of the history of mankind, would it be too bold to imagine, that all warm-blooded animals have arisen from one living filament, which THE GREAT FIRST CAUSE endued with animality, with the power of acquiring new parts attended with new propensities, directed by irritations,

sensations, volitions, and associations; and thus possessing the
faculty of continuing to improve by its own inherent activity, and
of delivering down those improvements by generation to its
posterity, world without end?[20]

Here, finally, we have a first attempt at thinking in terms of real antiquity. And there is a forerunner of his more famous relative's concept of
natural selection in stating that there are three great objects of desire
in every organism: the desire to reproduce, the desire to eat, and the
desire for security. Has anything changed from these in the lives of our
own species?

World without end, indeed. There will be an end to habitability
on this planet, but not for a very long time. Yet the most curious of
phrases in this prescient quote concerns the identity of "the great first
cause." Surely a divine Creator?

Erasmus Darwin, whose own genes would be passed on to his
grandson Charles, would be the first to experience the specific slings
and arrows that evolutionists of all scientific stripes have faced since.

The British press attacked Erasmus soon after his main publications. Yet that was not the worst of the calumny. For publishing this
revolutionary view of evolution, it was not Erasmus but his publisher
who was jailed! It was as if the British aristocracy, which then as now
controlled the media, quickly recognized the danger that any scientific treatise on evolution posed to the hardened English class system.

Thus, just as Stephen Jay Gould would be labeled a Marxist (which
in the 1960s and 1970s, amid the Cold War, was indeed disparaging)
for his scientific essays on evolutionary theory, so too was Erasmus
Darwin attacked by his own class immediately after his early provocative scientific and publishing feats.[21] By design he held back his most
cogent writings until after his death. That his grandson so delayed publication of his own even more antagonistic and revolutionary scientific
observations for so long was in no small way a result of knowing what

had happened to his grandfather: societal revulsion, excoriation, class dismissal.

While these eighteenth-century advances came two long centuries after the burning of Giordano Bruno for heresy in promoting anything other than an Earth-centered universe (and therefore a Christian God–centric universe), the societal memory of that horror remained strong even in Darwin's time. The stage had been set. Suggesting that evolutionary change could take place at all was perhaps even more revolutionary than Copernican theory. Naturalists were given enough social and material permission that they could look forward to knowledge that was not just a retread of the Greek and Roman civilizations. Linnaeus and Erasmus Darwin paved the way for Jean-Baptiste Lamarck and Charles Darwin, and the world was forever changed in cultures where religion steadily lost the power to stifle thought.

CHAPTER II

Lamarck to Darwin

F EW scientific revolutions have been more profound than that initiated by Charles Darwin in his classic book of 1859, *On the Origin of Species*.[1] It explained the basis for organic change through time: the process we now call evolution. Its importance was more than purely scientific, as it also became a rallying cry for and against change to the social fabric of the world's haves and have-nots soon after its publication. Today, we give Darwin primacy in authoring the overall scientific paradigm described as the theory of evolution, and its hold not only on natural science but also on many areas of social science continues. Yet Darwin was not the first naturalist to arrive at a comprehensive theory explaining organic change through time. He was not the first to attempt to account for change within an existing species, nor the first to explain the formation of entirely new species. Many nonscientists still believe that Darwin was the first "evolutionist." He certainly wasn't.

A half century before Darwin finally made public his momentous new theory, a retiring, hardworking French intellectual had walked a similar path and produced an entirely different and yet consistent set of ideas on how evolution occurs. Jean-Baptiste Lamarck was a soldier, biologist, academic, and aristocrat—and one of the great tragic giants in the history of science.[2] Although he was born of nobility, it was but an impoverished title that his parents bequeathed him. Yet he rose to prominence through his intelligence and drive, eventually becoming one of the most important curators in France's most important center of natural history, the Jardin des Plantes in Paris.

There might be no greater testament to his intelligence than his survival during the French Revolution, when so many others with noble titles intellectually lost their heads to anti- or pro-revolutionary fervor,

Jean-Baptiste-Pierre-Antoine de Monet Lamarck. Photogravure
after C. Thévenin, 1801. Wellcome Collection.

and then literally to the guillotine. Yet this revolution became a forging
fire to Lamarck's own intellectual turmoil, culminating in a truly revolutionary theory. In the first years of the nineteenth century, with Napo-
leon consolidating power, Charles Darwin still nearly a decade from
being born, and more than fifty years before the work of Gregor Mendel
that would discover the reality and form of genetics and how inheritance works, Lamarck published the first truly cohesive theory of organismal change through time. His theory of evolution explained that
some chemical force drove organisms up a ladder of complexity, and a
second environmental force adapted them to local environments
through use and disuse of characteristics, differentiating them from
other organisms. His own words present his case succinctly: "The environment affects the shape and organization of animals, that is to say that
when the environment becomes very different, it produces in the course
of time corresponding modifications in the shape and organization of
animals. It is true, if this statement were to be taken literally, I should

be convicted of an error; for, whatever the environment may do, it does not work any direct modification whatever in the shape and organization of animals."[3]

Revolutions, however, whether toppling governments or scientific paradigms, depend as much on luck and timing as substance. For his revolutionary theory about evolution, Lamarck had neither. His ill fortune was to coexist in the same institution with another intellectual giant, Georges Cuvier, the "father" of the field of comparative anatomy. Cuvier was a strong subscriber to what we now call intelligent design, as well as catastrophism (a theory that the earth has undergone periodic environmental calamities that exterminated all life), both antithetical to evolution as perceived by Lamarck.

Cuvier did everything he could to intellectually assassinate Lamarck,[4] and in fact he would have danced on his grave had not the aged, blind, and penniless Lamarck been "buried" in a shallow pit of lime rather than a grave at all, his flesh and bones rapidly consumed. Cuvier wrote the funeral oration, normally a paean to everything good that a fellow scientist had produced. Not in this case. Cuvier used embellished, stylish insult to bury the man and his theory. There are statues of both Cuvier and Lamarck in two of the most beautiful places in Paris: the Jardin du Luxembourg and the adjoining Jardin des Plantes. But Lamarck's statue is isolated from his scientific brethren. Lamarck and Cuvier had radically different ideas about their sciences and grandly different views of the most important studies that science should make concerning life. For Lamarck, it was about how species came about. But Cuvier, the father of mass-extinction research, was as concerned with death as with life. For those who study mass extinction, there is a monument built by Cuvier that is treated with far greater reverence than any statue, and was, in fact, its own kind of instrument that was used to demonstrate the reality of extinction.

In one of the halls near the edge of the park there is an incredible boneyard amassed by Cuvier, and it was this accumulation of skeletons, in this hall, that itself became a scientific tool. Prior to about 1800, there was no concept that past mass extinction events occurred at all. Cuvier was the first to draw attention to the concept of extinction by demonstrating that bones of large elephant-like animals found in Ice Age sedimentary deposits could not be assigned to any living elephant. After all, he had mounted the skeletons of all known elephants, and these new bones were decidedly elephant like, but not of any known kind. He deduced that these bones came from an entirely extinct species. Throughout his long career, he mounted the skeletons of as many mammals as the great French nation of the time could provide him from its worldwide efforts and trade and forced colonization of immense portions of the continents, and their presence allowed Cuvier to constantly see the anatomy of mammals on Earth—or at least those then alive that had been discovered by the far-ranging naturalists.

Skeleton by skeleton, Cuvier came to know the extant mammals probably better than anyone since, and certainly better than anyone before.[5] Thus, when fossilized bones of immense size were brought to him, he immediately recognized them as unfamiliar. By that time most of the continents had been explored to a greater or lesser extent, and while it was surely possible that many small mammals had yet to be discovered (as is the case today, but they are discovered at a very slow pace now), it was less possible there lived massive animals not yet encountered by mankind. Cuvier considered the size of the living animal from which these fossil bones came from, bones not yet entirely turned to stone, and determined they were not on par in age with fossils from the Paris Basin (now known to be of Cretaceous and Tertiary periods). These were bones of a relatively newly dead species. But where could such a creature live and not have been observed by this time? *"Nowhere!"* was Cuvier's answer. It must be *extinct*, he reasoned, and from this came the first concept of extinction in the modern sense. Cuvier, in the end, cared far more about the demise of species than their evolution.

With the demonstration that the extinction of single species was a reality, Cuvier then used what he knew of the fossil record to make a huge, though observationally sound, intuitive leap. He and his colleague Alexandre Brongniart used fieldwork to establish that invertebrate fossils found in Paris Basin strata that were higher in the piles of strata, and thus younger, showed to be of quite different faunas, and were different because many of the species in lower and thus older strata had disappeared. In some cases, not only a few disappeared in successively younger beds but most. Cuvier thus showed not only the reality of extinction but, using the fossil record, *mass* extinction.[6]

Lamarck certainly had his share of suppositions that were prescient as well.[7] One was that life came from inanimate matter, and there is no other way life could have come into existence on this planet unless it came here via panspermia, a transmission from elsewhere in the cosmos (with the highest probability being Mars), but that simply puts off the question by one more episode. But regarding the question of how the simple life that appeared from inanimate material underwent transitions to more complex life, such as animals and higher plants, Lamarck attributed this to the "striving for perfection" of what we call multicellular life, and in particular metazoans, or animals. To Lamarck, the perfection that life aimed for was humanity. This is pretty ironic for a man who was hounded throughout life and then even in death, who lived through a period when heads were lopped off in public squares. Perfection?

Yet it was Lamarck's conclusion that while this striving for perfection was a goal-oriented kind of evolution, he was not wrong in the overall idea of how it took place. He noted, and noted well, that animals live in environments that are always challenging, and that change, were it to take place, could reduce the challenge of staying alive. Thus, his most famous and derided example: that giraffes evolved longer necks because the living giraffes spent their lives stretching to get higher into trees where leafy food could be found. The longer neck was an adaption, and to Lamarck the inherent understanding that there is competition, and surely a sense that survival went to the longest-necked giraffes, would mean this trait was passed. The giraffe's striving for a longer neck and actually doing something about it physically became new and heritable. This is a good definition of heritable epigenetics, or, to give this genius his due, "neo-Lamarckism."

Lamarck was born too soon in so many ways, but one aspect of his time little commented on is how he was failed by the fossil record, or, actually, by the pre-geology naturalists who were fully cognizant of the fossil record. The ultimate irony is that, during his time, the world expert on sedimentary geology and interpretation of the fossil record was his rival and bête noir Georges Cuvier, and Cuvier, while making the break-through observations about the reality of extinction, also was guilty of an enormous fallacy—that no species could survive the mass extinctions.

Cuvier visited the many fossil beds in France. He certainly was aware of the rich ammonite faunas (ammonites being fossils of large size and striking morphology, thus being among the most obvious and well known of fossils from what we now call the Mesozoic era in the strata of France), and he observed the obvious change from the stratigraphi-cally highest strata that contained ammonites to the overlying beds, which while still having fossils no longer had ammonite fossils. From this came his understanding that the sedimentary record and the fossils it contained showed episodes of widespread death. Yet as perspicacious as he was, in this Cuvier made a fundamental mistake. He saw that the ammonites disappeared,[8] in what we now call the "K-T mass extinction," but he did not notice that many of the clams and snails found in the youngest ammonite-bearing strata in fact were also present in the oldest beds of the successive geological period, the Tertiary. To Cuvier, however, this changeover was evidence of an episode of environmental calamity that resulted in the death of *all* species, not just the ammo-nites. No survivors. And he held the same about the extinction of the mastodons. In his 1796 paper about fossil elephants that were dug out of beds near Paris, from what we now know as the Pleistocene epoch, he noted: "All of these facts, consistent among themselves, and not opposed by any report, seem to me to prove the existence of a world previous to ours, destroyed by some kind of catastrophe."[9]

But he also knew that some of these previous "worlds" giving up their fossils *had* to have been composed of environments very different from that of the Paris of his time. Specifically, in knowing that elephants of his time lived in warm climates, and certainly not in places such as Europe, he was seeing proof that there was change over time on Earth. These words from Darwin, written more than a half century later, were perhaps something that Cuvier intuited as well: "The theory of natural selection is grounded on the belief that each new variety, and ultimately each new species, is produced and maintained by having some advantage over those with whom it comes into competition; and the consequent extinction of less favoured forms almost inevitably follows."[10] Inherent is that conditions changed through time, forcing species to adapt or go extinct.

Cuvier butted heads intellectually with Lamarck about this, for Lamarck did not share this view of mass extinctions as catastrophes that wiped out all life, which in turn led to a repopulation with new organisms brought into existence by the Creator.

Cuvier danced around the concept of this Creator,[11] because he had two things to explain: the forces causing the mass extinctions and the forces bringing into existence the new animals and plants. Ultimately his Christianity led him to argue that the catastrophes marked in the geological record coincided with history as noted in the Bible. In spite of seeing the vast thickness of sedimentary rock in France and elsewhere, he fought against an Earth with an age counted in millions of years, which was another current belief of Lamarck and the French naturalist Étienne Geoffroy Saint-Hilaire, who not only argued with Cuvier about the reality of the total life extinctions but also that no animals or plants showed any evolutionary time while on

Earth. During his long life, Cuvier would not budge in his opposition to any evolutionary theory. He wrote: "This objection may appear strong to those who believe in the indefinite possibility of change of forms in organized bodies, and think that during a succession of ages, and by alterations of habitudes, all the species may change into each other, or one of them give birth to all the rest. Yet to these persons the following answer may be given from their own system: If the species have changed by degrees, as they assume, we ought to find traces of this gradual modification."[12]

This then became the rallying cry that exists to this day among creationists of all stripes, including those who use events in Earth history, such as the Cambrian explosion, to argue that there is a divine "watchmaker" tinkering with life on Earth. Where is the proof of change? asked Cuvier, and this is asked still.

Fortunately for science, Georges Cuvier was not the only savant of this time investigating the fossils and sedimentary rocks of France and nearby European countries. Lamarck was doing the same, exploring the sedimentary basins and their fossils scattered around Paris.[13] But Lamarck had an advantage over Cuvier in interpreting the fossils and the ancient environments that they must have come from.

Lamarck was interested in all aspects of the natural world, and it was not until he had made exhaustive studies of climate and weather, and then botany, that he turned to geology and paleontology. These latter fields were important in that Lamarck correctly noted that erosion into the sedimentary succession of the Paris Basin, mainly by rivers and streams, was the way in which the current geomorphology came about, and it was this work that brought the great English geologist Charles Lyell to begin to support and follow Lamarck's career,[14] since this was in line with what came to be known as the principle of uniformitarianism, which states that all geological processes occurring through time from the past to the present can be understood by looking at modern-day processes in nature.

This was an important refutation of a completely different world-forming view than that favored by Cuvier's overarching catastrophism theory.[15] Indeed, that Lamarck's work and its quite different view from the catastrophism of Cuvier was being taken seriously and publicly by an Englishman, amid the Napoleonic Wars no less, did not win Lamarck many French friends, and certainly must have enraged Cuvier, who was an ardent nationalist as well as being an intensely jealous man. Cuvier was ever eager for acclaim but little interested in supporting other naturalists, especially one actively communicating with the hated English. His jealousy[16] and hatred of Lamarck grew as the eighteenth century became the nineteenth.

Lamarck's ideas about fossils and the possibility of evolutionary change somehow related to environment first appeared at the end of a thick book named *Système des animaux sans vertèbres* in 1801, dealing with what was known about animals without backbones but also including other observations coming from Lamarck's work on living invertebrates and what he divined from studying extinct invertebrates. But as important as that book was, it was the 1809 publication by Lamarck of an immense tome named *Philosophie zoologique* (*Zoological Philosophy*)[17] explaining the natural history of animals that cemented his international reputation as a scholar and gave a platform for his maturing ideas about evolutionary change.

In this 1809 work, Lamarck explicitly defined his two "laws" of evolution.[18] The first law suggested that it was whether an animal used or disused some aspect of anatomy (including internal organs) that led to evolutionary change. Anatomical parts that were being used strengthened; those less used became weakened or disappeared entirely. The second law concerned the acquisition of new traits or modification of preexisting anatomy or organs in order to improve some aspect of the animal's life, stating that new traits accumulated during the lifetime of an organism, if useful and adaptive, would be passed on to new generations.

Lamarck, like Darwin after him, was obsessed with trying to understand how the world came to be populated with such an amazing variety, abundance, and diversity of life. And like Darwin, he wanted to know how life was able to change: to transform, as he put it. Also like Darwin, he came to his own great theory only after decades of patient study of all manners of both plants and animals. Unlike Darwin, however, Lamarck had to work for a living. He also buried four successive wives and several children, and he died in the abject misery of intellectual despair, knowing that the greatest biologists of his time rejected his views, and in so doing had not only engineered his descent into the penniless state of his last years but had succeeded in convincing the scientific world of his heresy and, more, his stupidity.

The dead never come back to life. Even the most deserving never get to see any sort of vindication; Van Gogh will never know the price of any of his paintings these days. And so too for Lamarck. Almost exactly two centuries after proposing the first internally consistent and possible theory of how evolution actually operates, if Lamarck could today have another day on Earth alive, we might believe that he would see that he is having his revenge against all the doubters.

Darwin's theory is taught in middle school science classes (if it is taught at all). Too many American schools bypass instruction in evolution rather than even "teach the controversy." Lamarck's contributions have been largely forgotten, and when they do appear in middle or high school science textbooks, they are used simply to show how wrong *anything* except Darwin's ideas about evolution were from the nineteenth century to today.

Yet, in this second decade of the twenty-first century, there is a "neo-Lamarckism" discussed among scientists. *Some* of the ideas that are descendants of the groping attempts by Lamarck and other naturalists of his time to describe evolutionary change are enjoying a comeback, because in the most powerful aspects they help explain observations and experiments in biology as well as observations coming from the fossil record.[19]

Great scientific theories start as simple discoveries that have great explanatory and predictive power. The theory of evolution is no different. Darwin's version began with what he saw as a simple law of nature, a law that could then explain a great deal of what had remained mysterious to all natural historians before him. He looked for reasons that could show how the enormous diversity of life now on Earth, as well as so many more species known from fossils, could come from some single, original cellular ancestor. But he was, by the mid-nineteenth century, tilling already well-trampled scientific ground.

Darwin was a geologist first of all, but so too was Lamarck. By Darwin's time, the nascent discipline of geology posited an Earth certainly old enough for life to have diversified to its present abundance. But it was the fossil record that gave the data used by both Lamarck and, after him, Darwin. Paleontology was the first and most important of the early contributors to the evolutionary theory of both.

CHAPTER III

From Darwin to the New
(Modern) Synthesis

T HE dichotomy of what is referred to as the theory of evolution is, in fact, both deceptively simple while at the same time as complicated as life itself. Life as we know it is composed of extraordinarily complex chemical assemblages that wall themselves off from a larger environment, extract energy from their surroundings, and, eventually, reproduce, making living copies of themselves. All who study life agree that these are *definitions* of what life does. But there is another property of life: that it cannot only reproduce but also *evolve*. Evolution is a way to improve efficiency, to withstand a changing environment, or even to outcompete rivals in competition for space or food.

Textbooks on evolution usually attribute Charles Darwin's observations of Galápagos Islands finches[1] as the primary reason that he came to propose the tenets about evolution through the process of natural selection, and thus our understanding of the theory. However, an understanding of evolutionary change had already been intuitive to most farmers for thousands of years before Darwin, since the domestication of animals and plants. English gentlemen of Darwin's time made their livelihood from agriculture. But because of the terrible English weather and short growing season, it was from sheep and cattle that much of their wealth came. No one raising livestock could help but directly observe evolutionary change in the directed formation of more productive livestock, and better dogs and horses to help raise them. To breed and make money from sheep, one certainly needs highly bred dogs. To supplement the wool and mutton meat, one needs pigs, chickens, and cattle, and selective breeding caused ever "better" strains by morphological

Portrait of Charles Darwin. Julia Margaret Cameron, ca. 1870. Library
of Congress Prints and Photographs Division.

change. But all of this change was observed by many inquiring minds,
including Darwin's.[2]

Breeding those animals over centuries produced a great deal of
insight for those English farmers observant enough to consider what
was happening in the brutal world of guiding the biological change of
formerly wild animals into creatures that could make the lord of the
manor a great deal of money from breeding them on his land. They had
to become tamer and larger in size (for the meat and, with sheep, the
wool) to increase the money to be made from them, and, if possible, they
needed to grow faster and produce more offspring than in nature. Dogs
were bred for helping humans get food but also for companionship.
They were bred to herd stupid, larger herbivores and eventually to fight
and kill—be it rats, predators, or humans—or to flush birds and then
retrieve them when shot. And to live sometimes solitary lives, no longer
members of a pack.

These "new" kinds of domesticated animals did not evolve by
normal, natural selection from their original, species-level origins; they

were *evolved* by humanity through a brutal kind of human-induced natural selection. Save those showing promise for the desired skills, the rest were killed at birth. It is no accident that some of the most important chapters in Darwin's masterpiece *On the Origin of Species* (itself one of the great misnamed works of literature, as it never overtly talks about the evolution at the species level of taxonomy) dealt with the domestication of animals.

Darwin arrived at his version of evolutionary theory over several decades of the nineteenth century. He came to his conclusions both by direct observation and also, as in most science, by borrowing facts and conclusions from scientists who came before him. One such scientific predecessor who was crucial to Darwin was the influential English savant Thomas Malthus, who presciently wrote about the dangers of rapidly growing population sizes.[3] Malthus wrote specifically about too many humans, but Darwin realized that the danger of large populations was relevant to *all* living things, not just humans.

Malthus did not write in a vacuum, but he too came to his conclusions based partly on prior observations. For Malthus, a major influence was the polymath American Benjamin Franklin, who preceded Malthus in warning about the effects of enlarging *human* populations, with Franklin focused on the American colonies. Yet it was Malthus who most powerfully understood what our current society is just truly awakening to now: that in spite of our amazing ability to increase crop and food production through science, there is a finite limit of food supply for humanity, one that we will certainly be affected by this century, given that global warming is melting the Greenland and Antarctic ice sheets, causing sea level rise to potentially threaten low-level agriculture around the world. Sea level rise (a significant amount of global food is now raised at elevations that will be occluded by sea level rise in this century and the next) coupled with the reduction of crop yield in an ever-higher carbon dioxide atmosphere could make these words of Malthus prescient:

The power of population is so superior to the power of the earth to produce subsistence for man, that premature death must in some shape or other visit the human race. The vices of mankind are active and able ministers of depopulation. They are the precursors in the great army of destruction, and often finish the dreadful work themselves. But should they fail in this war of extermination, sickly seasons, epidemics, pestilence, and plague advance in terrific array, and sweep off their thousands and tens of thousands. Should success be still incomplete, gigantic inevitable famine stalks in the rear, and with one mighty blow levels the population with the food of the world.[4]

These words affected Darwin immensely. His was an intellect that could assimilate the reality of this argument not only for humanity but for any population of animals, and there is an argument to be made that following his return from the epic voyage of the *Beagle*, the increasingly reclusive Darwin was more interested in the animal kingdom than the human kingdom. But in any event, the work of Malthus was critical in the intellectual development of Darwin's great breakthrough. Or three breakthroughs.

Darwinism is based on three readily understandable "propositions," as they became known. The first is that any population of a given species will produce more individuals than can be supported by the resources of the environment. The second proposition is that those with anatomical or physiological characteristics rendering them less susceptible to death—be it from starvation by competition from others of its kind, or by predation by predators of any species, or by physical effects—will preferentially survive, and in so doing will ultimately produce more offspring than individuals with less advantageous traits. Darwin's famous "survival of the fittest" refers to fitness as the ultimate prize in biology: living long enough to reproduce and then successfully having offspring. Since most organisms are killed early in life in nature, this is no mean feat. The third proposition is that those characteristics allowing survival will be passed on to the progeny. The traits that led to survival (sometimes by luck alone, more often not) are then possibly lending success to the next generation.

Darwin is given great credit in our time for the third proposition, the one overtly describing heredity, as there was no rigorous scientific basis in his time to what we now call genetics. Darwin conceived of evolution by natural selection without knowing the makeup of what we now call "genes." But he certainly understood that they must exist. Darwin posited that the factors enabling survival of an individual are "selected for." The discovery of DNA was a century in the future when he finally finished his great literary masterpiece. But even the least observant human would know that children take after their biological parents and grandparents. In nature, among the multiplicity of the new generation of the many parents making up a population, some would live, some would die. Darwin's genius was in his prescient understanding, rare for his time, of how *long* geologic time really was. Darwin believed that it was the slow rise and fall of populations based on natural selections over vast periods of time that ultimately were needed to produce evolutionary change. Even a millennium was but a tiny interval compared to the vastness of time that populations of most species lived through. No mere agriculturalist would ever arrive at that conclusion. Ultimately, the theories of evolution needed geology to be made real, and geology, during Darwin's life, had come into its own as a science that was first and foremost concerned with time and its measure.

DARWIN VS. LAMARCK

In an odd twist of symmetry, the year 2009 was the 150th anniversary of *On the Origin of Species*, and it was the two hundredth year since Darwin's birth in 1809. But it was also the two hundredth anniversary of Jean-Baptiste Lamarck's *Philosophie zoologique*, in which Lamarck cogently described his own theory about evolutionary change in organisms: that it is driven by beneficial "phenotypic" (such as specific morphology) changes that were not randomly acquired but came about through an intersection of the organism with some aspect, or more commonly some challenge, of that organism's environment—and that this intersection not only changed the life of the organism from that time onward, it was also heritable. But fifty years later Darwin championed something different: that the changes of an organism were random, and not directed, and that it was simply the sum of morphology in a large population that was worked on by his newly defined "natural selection."[5]

While Darwin either ignored or refuted the Lamarckian postulates in the first two versions of his book, he updated his initial reluctance to embrace any aspect of Lamarckism in later editions.[6] Over time, Darwin warmed to some of Lamarck's beliefs, while at the same time continuing to distance himself from giving much credit to his poor dead predecessor. For instance, starting with the third edition of *On the Origin of Species*, Darwin recanted and began to include aspects of Lamarckism, the most important being Lamarck's concept of inheritance of acquired characters/characteristics (IAC).[7]

Beginning with the third edition, Darwin accepted the possibility of this IAC, but he considered it minor in importance compared to his favored mechanism: random, undirected variation. But as he continued to update his books, Darwin gave ever more credence to IAC, and in the end suffered critical attacks for this change. Criticism against those (the epigeneticists) favoring IAC as being both real and important in evolutionary change continue to this day, for IAC is the heart of heritable epigenetics.

There is much not to like in Lamarck's conception of evolution. He did not believe in extinction but instead believed that species transformed from one to another, and did so because they followed what he thought of as a drive toward perfection, generation by generation. Yet,

new discoveries require a thorough reconsideration of the century or
more of automatic rejection of Lamarckian precepts.

TEST OF TIME

Darwin's theory of evolution has survived every scientific challenge
since the first publication of *On the Origin of Species*—until this
century, that is. Here is Darwin's theory in concise form:

1. There is a pattern of characters encoded in each organism
   in structures called genes.
2. This pattern is copied and passed on to offspring.
3. The copying is never perfect: Variations arise through
   errors in copying or through random (not directed)
   mutations. This produces variation. Even greater variation
   is introduced through sexual reproduction.
4. The variant members compete with each other, for more
   offspring are produced than can survive.
5. There is a multifaceted environment that makes some of
   the variants more successful than others.
6. The individuals that survive and go on to reproduce, or who
   reproduce the most, are those with the most favorable
   variations. They are thus *naturally selected*.

A key point in this is that natural selection is not forward-looking:
Each generation is always adapted to the environment of its parents
(as seen in insects such as cicadas, which can spend decades as buried
larva, before finally emerging into a world that may or may not have
conditions that favored their parents). Evolution does not adapt an
organism to possible future conditions, only to conditions that its parents
experienced.

Today, a revival of Lamarckism (the previously mentioned "neo-Lamarckism") is causing many biologists to admit that a major revolution in evolutionary theory might be at hand. The cause? The discovery that small molecules attaching to the DNA molecules that hold genes along their enormous length are capable of causing biological change similar to what mutations can do, but faster. But mutations are simply a change in the letters of the genetic code making up some part of a gene. This new process, a kind of epigenetics, leaves that genetic code untouched; the various nucleotide combinations that call for specific amino acids, that themselves are then sewn together into myriad proteins that allow the business of life, are still in their original order. But the small molecules that sometimes attach to them change their actions. It is as if the genetic code itself has been changed. And the actual biological change has occurred because of an action taking place during the organism's life, in a manner described by Lamarck.

A major tenet of Darwinism is that no trait acquired during the lifetime of an individual will have any genetic effect on that individual, even though such traits might help an individual's survival. Let's say the beak of some finch in the Galápagos Islands was broken through accident and the new jagged edge helped that individual eat thorny cactus more efficiently, and that this new kind of beak was then passed on to the offspring of this lucky bird so that all of its new chicks had this same kind of beak. This is absurd. But events occurring during the life of the parent can cause change, if not so directly as in this example.

Darwin never did know the identity of the mechanism of heredity, although he knew that heredity had to exist. But he certainly knew that the process described by Lamarck (like all young English gentlemen of his time, Darwin read French) as the "inheritance of acquired characteristics" was at the heart of the only other scientific theory of his time, or before his time, that tried to explain evolutionary change. By Darwin's time, this mechanism for inheriting an acquired characteristic had another name, one that was a name of scorn: Lamarckism. Another quote from Lamarck puts this in context:

> Do we not therefore perceive that by the action of the laws of organization . . . nature has in favorable times, places, and climates multiplied her first germs of animality, given place to developments of their organizations, and increased and diversified their organs? Then . . . aided by much time and by a slow but constant diversity of circumstances, she has gradually brought about in this respect the state of things which we now observe. How grand is this consideration, and especially how remote is it from all that is generally thought on this subject?[8]

While a timid man, the Charles Darwin who lived years after the publication of *On the Origin of Species* knew that his great theory, if it was to be accepted, needed as much PR help as it did further scientific proof. He was certainly salesman enough to know that he had to bury the competition. Thomas Henry Huxley, among others, attacked the by then long-dead Lamarck.[9] Inheritance of an acquired characteristic or trait? Impossible. Until, that is, it became known that it is not just possible but common, and in so happening (it appears to me), it has had, currently is, and will continue to produce enormous social and biological consequences for our own species.

Darwin hoped that the history written in the fossil record would sooner or later support his contention about his theory of evolution: that change came about in small increments. To paraphase Darwin, "From the beginning of life on earth there was a connection so close and intimate that, if the entire record could be obtained, a perfect chain of life from the lowest organism to the highest would be established."[10]

He also explicitly stated how evolution from one species to the next would take place: that (1) new species arise by the transformation of an ancestral population; that (2) transformation is even and slow; that (3) transformation involves most of all of the ancestral population; and that (4) transformation occurs over most or all of the ancestral population's geographic range. Darwin wrote of his expectation of the fossil record that it should depict a continuous and observable (as fossils) lineage, as he noted, "as by this theory, innumerable transitional forms must have existed, why do we not find them embedded in countless numbers in the crust of the earth?"[11] It was due to imperfections in the geological record.

Darwin found solace in inserting in his own written statement an old quote from Carolus Linnaeus:

> As natural selection acts solely by accumulating slight, successive, favorable variations, it can produce no great or sudden modification; it can act only by very short and slow steps. Hence the canon of *"Natura non facit saltum,"* which every fresh addition to our knowledge tends to make more strictly correct, is on this theory simply intelligible.[12]

Natura non facit saltum: "Nature makes no leap," meaning that evolution took place slowly and gradually. This was Darwin's core belief. And yet that is not how the fossil record works. The fossil record shows more "leaps" than not in species.

Nature makes no leap. Yet Darwin saw the leaps made by the many domesticated animals on his large farm. Even on a timescale only in years rather than millions of them, many dogs, pigs, cows, and chickens (especially chickens!) show "leaps." *Natura facit saltum.* However, beyond domestic animals, the scaling is wrong to really understand speciation. Microbial speciation can take place on the timescale of days or weeks, apparently. For wild animals it is longer. But the nature of the fossil record, if analogized to a book, is that its pages are indeed paper thin, yet temporally thick: The pages are strata. And strata are the net result of soft sediments, which become buried, compressed, lithified. Yet little was known about how long it took to create each stratum, to fill it with fossils and thus "print" the data of the fossil record. In some deepwater deposits there seem to be no more than one complete stratum every twenty thousand years. For the tiny plants called coccoliths (whose skeletons make up chalk, and whose self-same skeletons contain enough morphology to recognize differences in species as readily as we can distinguish human fingerprints), the time to produce a new species is probably far less than twenty thousand years. Each stratum can thus have an entirely different species in it, with none of the intermediates visible. These were the intermediates demanded by Darwin. But sedimentary processes, not evolution, failed him. We know now that sedimentation is not fast enough to catch speciation events, and this is what initiated the fallacious mendacity of creationism.

THE TWENTIETH CENTURY

Charles Darwin, in edition after edition of his great masterpiece, railed against the fossil record: The problem was not his theory but the fossil record itself. Because of this, paleontology became an ever-greater embarrassment to the Keepers of Evolutionary Theory. By the 1940s and '50s this embarrassment only heightened. Yet data are data; it is the interpretation that changed. By the mid-twentieth century, the problem posed by fossils was so acute that it could no longer be ignored: The fossil record, even with a century of collecting after Darwin, still did not support Darwinian views of how evolution took place.

The greatest twentieth-century paleontologist, George Gaylord Simpson, in midcentury had to admit to a reality of the fossil record: "It remains true, as every paleontologist knows, that most new species, genera, and families, and that nearly all new categories above the level of families, appear in the record suddenly and are not led up to by known, gradual, completely continuous transitional sequences."[13]

Yet while paleontology seemed to argue against Darwin,[14] other fields supported him. From the 1930s to the 1950s, the dominant paradigm of Darwinian evolution (slow gradual change caused by single, random mutations over long periods of time) was seemingly strengthened by the discoveries of genetics: The works and the analysis of the evolutionists Ronald Fisher and Theodosius Dobzhansky reinforced the paradigm of evolution based on gradually changing gene frequencies. It combined natural selection and the then rapidly evolving field of genetics and breakthroughs in developmental biology into a consensus that became known as the "modern synthesis," sometimes called the "new synthesis."

The modern synthesis allowed the evolutionary process to be described mathematically for the first time, as frequencies of genetic variants in a population change over time—as, for instance, in the spread of genetic resistance to the *Myxoma* virus in rabbits.[15] Yet in the 1960s came the new concept of "allopatric speciation":[16] that new species form not by gradual transformation but by a small number of the mother species becoming geographically isolated and then rapidly adapting to new conditions. After enough time, if the small, new "founder" population and original mother population from which it separated are reunited, they can no longer interbreed. By that most rigid of definition of a species, they are now separate species. Harvard's Ernst Mayr, who would become a grand magister of the modern synthesis, described the concept of allopatric speciation as follows:

> The major novelty of my theory was its claim that the most rapid evolutionary change does not occur in widespread, populous species, as claimed by most geneticists, but in small founder populations . . . As a consequence, geneticists described evolution simply as a change in gene frequencies in populations, totally ignoring the fact that evolution consists of the two simultaneous but quite separate phenomena of adaptation and diversification.[17]

Thus it was in the middle part of the twentieth century that the importance and influence of paleontology in the "high tables" of evolutionary theory diminished, finally bottoming out with the statement by the Nobel Prize–winning physicist Luis Alvarez, who in frustration with the paleontologists of the early 1980s labeled them as mere "stamp collectors."[18]

As noted earlier, the fossil record is too coarse to demonstrate speciation. Mostly, evolutionary change happened so quickly that the fossil record sees only the ancestor and the descendant. If, for instance, some species of frog had a small population that became isolated from its larger mother population, and this small number of separated frogs found themselves in a new environment that favored very different traits than was important from the region where their species itself was first formed, what is the chance that this small, separated environment will itself leave behind a fossil record? And if later members of this new species made their way back to where the original population still lived and reproduced to the point where they left fossils, there would be no record of evolutionary intermediates.

But there is another case that can be made. What if the separated population of frogs found themselves in an environment that was favorable to preserving their dead as fossils, accumulating in strata over time? And what if these strata formed episodically, not constantly? If the generational length in years is the ability of dead frogs to be preserved by the occasionally produced strata, where there might be time for fifty or a hundred generations before a new fossil-bearing bed is laid down, what we will see is the "sudden" appearance of the frogs from fifty generations later. Now add to this the realization that evolutionary change through epigenetic means appears to be able to occur orders of magnitude *faster* than the rates of change caused by Darwinian random, chance mutations. The "appearance" of the new species as fossils will seem even more instantaneous.

Thus many or most species appear seemingly instantaneously, at least as fossils. Species so changed from what seemed to be their direct ancestors that far more than a simple, single mutation was required. Under the modern synthesis, mutations are random. They are not directed. Yet, in case after case, species suddenly appear in the fossil record, even where there is rapid sedimentation and the "insensibly graded series" suggested by Darwin ought to be visible. Even in these cases, such as in deep-sea deposits filled with the fossils of microscopic calcareous or siliceous plankton, the fossil record suggests something that seems impossible based on the modern synthesis.

But a seeming solution came as the case for allopatric speciation strengthened in the 1970s and '80s. The hypothesis that speciation took place in separated and small populations was important in the interpretation of the fossil record, and once the primed and inquisitive minds came along, out came the most important contribution to evolutionary thought in a century. This became the scientific basis for the revolutionary hypothesis by Niles Eldredge and Stephen Jay Gould named "punctuated equilibria,"[19] or "punk eek," as it affectionately came to be known in the late 1970s. The case of the frogs imagined in the paragraphs above is an example: a tiny population cut off from the larger mother population. The probability of the tiny founder population preserving into the fossil record would be infinitesimal.

Eldredge and Gould melded this concept to the nature of the fossil record. Here is a quote from one of their earliest papers, from 1977: "The theory of allopatric (or geographic) speciation suggests a different interpretation (from that of Darwin) of paleontological data. If new species arise very rapidly in small, peripherally isolated local populations, then the great expectation of insensibly graded fossil sequences is a chimera. A new species does not evolve in the area of its ancestors."[20]

Further, Gould had this to say about the ability of the fossil record and evolution to leave evidence of the "insensibly graded series" of transitional fossils demanded by Darwin in support of his theory: "The extreme rarity of transitional forms in the fossil record persists as the trade secret of paleontology. The evolutionary trees that adorn our textbooks have data only at the tips and nodes of their branches; the rest is inference, however reasonable, not the evidence of fossils . . . All paleontologists know that the fossil record contains precious little in the way of intermediate forms; transitions between major groups are characteristically abrupt."[21]

And so, by the end of the twentieth century, evolutionists tried to sum up the processes of new species formation:

1. New species arise by the splitting of lineages.
2. New species develop rapidly.
3. A small subpopulation of the ancestral form gives rise to the new species.
4. New species originate in a very small part of the ancestral species' geographic extent—in an isolated area at the periphery of the range.

These four statements again entail two important consequences: (1) In any local section of fossils bearing rocks containing the ancestral species, the fossil record for the descendant's origin should consist of a sharp morphological break between the two forms. The break marks the migration of the ancestral range. (2) Many breaks in the fossil record are real: They express the way in which evolution occurs, not the fragments of an imperfect record.

SUBSEQUENT TO THE MODERN SYNTHESIS: DARWINIAN EVOLUTION UNDER ATTACK IN THE TWENTY-FIRST CENTURY

At the turn of the century, the modern synthesis still retained the theories that new sources of morphological and/or physiological variation arise through *random genetic mutation*; that inheritance from genera-tion to generation occurs *only* through DNA being passed to the next generation; and that natural selection is the sole cause of adaptation. But about the same time, other voices were being raised in increasing dissent. Their points were a shot across the bow of these traditional acceptances. Most important of the dissenting views was that there remained important "missing pieces" to evolutionary theory. Chief among these was that ascribing all evolutionary change as entirely "gene-centric" was a mistake, as was the evolutionary establishment's dictate that there was no so-called soft inheritance, a term meant to encapsulate the possibility of heredity by modifications to genes without the genes themselves being rewritten from their original genetic code.

The new views were that too little attention had been paid to how variation in members of a species could also come from differences during an organism's development, from fertilization to birth; that too little attention had been paid to how vagaries of the environment expe-rienced and lived in by an individual could affect its ultimate biological makeup (from morphology to physiology to behavior) but especially the increasingly observed phenomena where organisms were transmitting more than genes across generations. In the twentieth-century view, these phenomena are just outcomes of evolution.[22] To the biologists who were increasingly calling themselves "epigeneticists" in the late twen-tieth and early twenty-first centuries, they are also important causes of evolution.[23]

Biologists use the term *plasticity* to describe morphological or other traits that are highly variable, be it in shape, physiological aspects, or even behavior: Plastic traits are labile (or variable) and generally show a wide variety of different variations within a single trait. One such trait is hair color in dogs: Within any litter, there can be a wide variety of fur color among the puppies. An increasing understanding is that a high plasticity may increase the survivability of the species possessing the plastic traits. But the new twist is that plasticity not only allows organisms to *cope in new environmental conditions* but to *generate traits* that are well suited to them when confronted by a radically new kind of environment or environmental condition. In this view, it is the trait that comes first; genes that "cement" the trait in heritable fashion only occur afterward, and this may not happen until several generations later. New adaptations or traits can thus be environmentally induced, but once in place these adaptations may then allow colonization of some other kind of environment, and in so doing cause isolation of a small population that may become a new species.

A consequence of the Darwinian model is that every individual organism that is born is a product of parents successfully breeding because of traits (from genes) that led to their survival *under the conditions in which their own parents lived*. Darwinian evolution is thus backward-looking. In environments that are changing rapidly, those species with a long developmental period and slow growth to maturity may find themselves in a radically different environment than was present during the lives of their parents. Think of the aforementioned cicadas, insects that as juveniles spend decades underground, to then emerge and breed after a long Rip Van Winkle–developmental period. In those decades underground, the environment they emerge into might be different indeed. But their survivability can be enhanced through morphological and physiological plasticity, and it is the new environment, not the genetic adaptation to the old world of their parents, that can be brought into play through epigenetics.

The current state of play is the degree to which the field of evolutionary theory needs to be amended. That is indeterminate until years go by. But as historians are wont to say, history is written by the victors. Whether epigenetic findings are indeed revolutionary or are just another addition to the evolutionary edifice is just semantics. But for my own field, paleontology and paleobiology, the epigenetic self-described revolution gives us a totally new kind of time machine with which to plumb and interpret the long dead.

CHAPTER IV

Epigenetics and the Newer Synthesis

L ET us begin this chapter, on the modern understanding of epigenetics, in a way similar to the preface of the book, by looking at evolution in the externally shelled cephalopods I have devoted a lifetime of study to. In this case it is not fossils but the modern nautilus species that are now beginning to go extinct in the regions where they are fished. As a caveat, it seems clear that one trait no animal or plant on Earth can bear with risk is being "attractive" to humans. From feathers to rare plants, and from butterflies to nautiluses and so many other beautiful seashells, to be "collectable" is to be endangered.

In 2012, I ran a trip sampling the nautilus populations along Australia's Great Barrier Reef explicitly to see if nautiluses living on marine protected areas of the reef are as rare as from places where they are fished for their pretty shells (such as in the Philippines and Indonesia). Work along the Great Barrier Reef in the 1990s had shown that two different and accepted species are present. One, *Nautilus pompilius*, is the most widespread of all the nautiluses across their vast Pacific and Indian Ocean range. The second, *Nautilus stenomphalus*, is found only on the Great Barrier Reef. It differs from the more common *N. pompilius* in having a hole right at the center of its shell. (In *N. pompilius*, there is a thick calcareous plug.) There are also marked differences in shell coloration and pattern of stripes on the shell. But when the Australian species was first brought up from its thousand-foot habitat alive, in the late twentieth century, scientists were astonished to find that *N. stenomphalus* has markedly different anatomy as well on its thick "hood," a large fleshy area that protects the interior guts and other anatomical soft parts when the animal pulls into its shell. In *N. pompilius* the hood is covered with low bumps of flesh, like warts.

Meanwhile the *N. stenomphalus* hood is covered with a forest of brushy projections that rise above the hood like a thick carpet of twiggy moss, or tiny trees of flesh; the coloration of the hood is also radically different.[1]

The 2012 trip was to sample the DNA of the two "species" as well as to better understand how many nautiluses live on a given area of seafloor. We caught thirty nautiluses over nine days, snipped off a one-millimeter-long tip of one of each nautilus's ninety tentacles, and returned all back to their habitats alive (if cranky). All the samples were later analyzed in the large machines that read DNA sequences, and to our complete surprise we found that the DNA of *N. pompilius* and the morphologically different *N. stenomphalus* was identical.[2] No genetic difference, yet radically different morphology. The best way to interpret this is to go back to one of the most useful analogies in evolution: of a ball rolling down a slope composed of many gullies. Which gully the ball rolls down (corresponding to the ultimate anatomy or "phenotype" of the grown animal) is controlled by the direction of the push of the ball. In evolution, the ultimate morphological fate of an organism is caused by some aspect of the environment the organism is exposed to early in life—or, in the case of the nautiluses, while they slowly develop in their large egg over the course of an entire year before hatching. Perhaps it is a difference in temperature. Perhaps it is forces that the embryo encounters prehatching, or when newly hatched, the small nautiluses (one inch in diameter, with eight complete chambers) find different food, or perhaps they are attacked and survive, i.e., have two different kinds of predators. That's why *N. pompilius* and *N. stenomphalus* are not two species. They are a single species with epigenetic forces leading to the radically different shell and soft parts. Increasingly it appears that perhaps there are fewer, not more, species on Earth than science has defined.

More and more, biologists are discovering that organisms thought to be different species are, in fact, but one. A recent example is that the formerly accepted two species of giant North American mammoths (the Columbian mammoth and the woolly mammoth) were genetically the same but the two had phenotypes determined by environment.[3]

THE THIRD EPOCH: EPIGENETICS ADDED

The development of evolutionary theory encompassed three major stages: Lamarckism of the late eighteenth century gave way to Darwinism in the nineteenth century, and then modern synthesis modified the Darwinian set of theories in the twentieth century with the additions of paleontology, genetics, and the results from molecular biology (such as the discovery of DNA). The additions enlarged and gave ever more nuance, as well as explanatory power, to what was and is called the theory of evolution. Yet, even with these milestones of discovery, major questions remained unanswered, most notably about instances from the fossil record of evolutionary change that appeared to happen without intermediaries.[4] The lack of the mythic "missing links" gave ever more ammunition to those invoking the supernatural, which is, in fact, what many major religions rely on.

But our twenty-first-century discoveries coming from epigenetics again required addition or modification to evolutionary theory.[5] In some quarters, the discoveries coming with ever more rapidity from those studying epigenetic processes were claimed to be no less than a "scientific revolution." Others were less sanguine. Yet what no one denied was that the epigenetic discoveries were important regardless of whether they were viewed as revolutionary. Much of the dispute, however, arose from the quite variable use of the word *epigenetics* itself.

There are many conflicting uses of the term *epigenetics*, and this as much as anything has led to great dissension among and between scientists, as well as between scientists and science journalists. This is not an isolated incident: There are many cases in science where specific terms are used in quite different contexts, by different scientists, where the same word takes on disparate meanings; as a consequence, confusion can arise. In the past decade alone, there have been an increasing number of books, popular articles, and scientific reviews concerning epigenetics and in them there has been a diversity of meanings and ways that the word has been used. (And, according to many critics, overused.)

The origin of the word comes from British biologist Conrad Waddington,[6] for whom *epigenetics* was the study of how "genotype" (the sum of genes contained by an organism) is translated into "phenotype," the actual physical manifestation of the organism, as well as its various and specific chemical properties and productions and, as we increasingly know, its behaviors. But to other scientists, there is a far more specific sense to the term: *Epigenetics* is the study of heritable gene functions *that are passed on* from one reproducing cell to another, be that to a somatic (body) cell or to a germ cell (sperm or ovum), which does not involve a change to the original DNA *sequence*. It is the latter case that can lead to major evolutionary change. During the process of "meiosis," the replication of cells (sperm and eggs) in sexually reproducing organisms, information is put into the sperm or ovum that will, like some exotic secret writing, become readable only after fertilization.

Epigenetics (or heritable epigenetics, or neo-Lamarckism) is a series of different processes that can cause evolutionary changes as well as dictate how organisms develop from a single fertilized egg (in the case of sexually reproducing organisms, at least) to what we look like as adults. Some say it's just a minor tweak of already understood processes and that it's of little importance in the broader scheme of evolutionary change or the past or even future history of life.[7] But to others epigenetics, while still poorly understood, is potentially of far greater importance than mainstream evolutionary theory, and mainstream evolutionists have heretofore accepted that. To a few, its ongoing discovery is causing an unfolding scientific revolution. But the discoveries have not happened evenly among the many fields within what we call "biology." The great breakthroughs have mainly been studies looking at cells, and the molecules within cells, including DNA and RNA and other aspects of genetics. But to date there has been little if any progress in tying epigenetic change to the many events evidenced by fossils and the fossil record.

In genetics, genes are disrupted by mutations and permanently changed. Epigenetic effects take place when single genes along a long strand of DNA become "polluted" with very small molecules, which each attach to only a single small site along the long DNA molecule. This can cause a gene that was actively in use—such as one dictating the production of a specific protein—to become blocked from its normal activity. That protein is no longer made. But sometimes one single such block can affect the normal operating of hundreds of genes, such as when a master control gene (called a *Hox* gene) is inadvertently turned off. Because *Hox* genes control *hundreds* of other genes by telling them when and where to turn on and off, a single epigenetic change to that gene now affects a vast number of other genes. *Hox* genes dictate the building of organs, limbs, skin, and every part of a developing organism. Causing a *Hox* gene to turn off can have profound biological effects far greater than any single mutation. In this way, epigenetic change can radically and quickly transform the anatomy of an organism—for better or worse.

In epigenetics, genes that are inactive (silent) thus can be awakened and begin causing biological effects in an organism by environmental stimuli that would not happen if those environmental stimuli were absent. They are not necessarily permanent changes: The small attaching molecules are not permanently welded in place; DNA has long ago evolved the means to repair itself, including the removal of these bad molecules. Thus, in most cases epigenetic changes that affect us have no effect on our offspring. But sometimes these epigenetic changes do get passed on through eggs and sperm.

The study of epigenetics really comes down to observing two types of epigenetic changes. The first type of changes are the "normal" epigenetic changes that organisms go through, honed by natural selection. For instance, every cell in our bodies contains all the necessary information to become one of the many *specific* kinds of cells necessary to keep us alive, such as the nerve cells, muscle cells, and the many other highly specialized cell types that are necessary for living. Every cell contains the DNA information to become any or all. But it does. But they do not. The science involved in epigenetics looks to understand how it is that a specific cell at a specific time in a specific anatomical place "knows" how to change into something quite different according to time, place, and function. But the changes are "foreseen" by the organism and beneficial.

The second kind of epigenetic change causes unforeseen modification to an organism without altering the genetic coding for specific genes, but it also passes on these changes. It can cause change ranging from minor to profound, and can be heritable. "Lamarckian" change is where something encountered in its environment, and not necessarily expected in the life of an organism, causes chemical changes to the DNA through the addition of tiny molecules, or through a shape change of the scaffolding that holds the twisted DNA molecules in specific shapes. Other kinds of epigenetic change can also be caused by the actions of small RNA molecules responding to some kind of external environmental change.

Each of these can change how genes act by turning genes on or off. This can include some of the most important genes for our lives, the ones that affect our behavior through the rate at which hormones dictating emotions are regulated and supplied.

Here is a fuller description of the most important identified means by which epigenetic changes are produced:

Methylation is the addition of very short chains of carbon, oxygen, and hydrogen to particular nucleotides in DNA, which typically silences gene activity.

Histone modification involves the chemicals (histones) that serve like support structures for a DNA molecule. They can cause the shape of the DNA to change by making it more or less packed on itself. When they are modified by the addition of one of several small chemical molecules (again, a methyl molecule, which is the tiny molecule with a single carbon atom accompanied by hydrogen atom), as well as additional small chemical groups composed of only a few atoms, they are added on to the much larger histones, thus changing the overall shape of this chemical "scaffolding" that holds the DNA molecule within the cell. When so packed, the DNA is harder to get to by the small molecules of RNA trying to read the code, and they go to the cell's protein factories, such as ribosomes, where proteins called for by the DNA are actually built.

A third kind of change is caused by tiny RNA molecules (RNAi) affecting the chromatin (the histones) described above. In fact, a diverse assemblage of different-length RNA molecules are now known to be regulators of gene expression, as well as being used in genome defense against foreign genetic elements such as attacks on a cell by a virus. Small RNAs modify the shape of the chromatin structure and can stop (silence) the process known as *transcription*, where a gene dictates which protein should be built.

Sometimes an epigenetic change causes a protein not to be made. Sometimes it causes the making of a new protein that would not otherwise occur. Sometimes, and most important, it causes a regulator gene (essentially the "general contractor" coordinating all of the cells on the body's busy construction projects) to walk off the job entirely. This causes huge changes far beyond what any single mutation could do. Such changes affecting an individual can then be passed to the next generation. The methyl molecules are not physically passed on to the next generation, but the propensity for them to attach in the same places in an entirely new life-form (a next-generation life-form) is. This methylation is caused by sudden traumas to the body, such as poisoning, fear, famine, and near-death experience. None of these events come from small methyl molecules, but they cause small methyl molecules already in the body to swarm onto the entire DNA in the body at specific and crucial sites. These acts can have an effect not only on a person's DNA but on the DNA of their offspring. The dawning view is that we can pass on the physical and biological effects of our good or bad habits and even the mental states acquired during our lives.

This is a stark change from the theory of evolution through natural selection. Heritable epigenetics is not a slow, thousand-year process. These changes can happen in minutes. A random hit to the head by an enraged lover. A sick, sexually abusive parent. Breathing in toxic fumes. Coming to God in religious ecstasy. All can change us, and possibly change our children as a consequence.

In heritable epigenetics, we pass on the same genome, but one marked (*mark* is the formal term for the place that a methyl molecule attaches to one nucleotide, a rung in the ladder of DNA) in such a way that the new organism soon has its own DNA swarmed by these new (and usually unwelcome) additions riding on the chromosomes. The genotype is not changed, but the genes carrying the new, sucker-like methyl molecules change the workings of the organism to something new, such as the production (or lack thereof) of chemicals necessary for our good health, or for how some part of the body is produced. Thus, the young of an epigenetically modified parent can be radically different in phenotype from the parent. Phenotype is the physical manifestation of genotype, such as hair and eye color or body dimensions in a human—or of IQ and brain functioning. Sometimes these changes allow the young organism to deal with environments that were intolerable to the parents. Sometimes these changes rapidly create new species. But sometimes the consequences can be fatal and the changes can be passed on to yet a subsequent generation. In other words, a young child could suffer from the sins of a grandfather.

New experiments raise scientifically and morally important questions for our own species and our future evolution. Of all the aspects concerning epigenetics, none is more controversial than heritable epigenetics. The way that epigenetically produced changes can be passed on can be thought of as both "direct" and "indirect"; neither of these are formal definitions but commonsense conclusions from the scientific literature.

In the direct method, the places on DNA where methylation has occurred are marked in such a way that they are passed on to the fertilized egg. As the new organism develops, these sites on the new genome again become methylated. The critical point is that if "plastic" changes caused by environmental effects on morphology (phenotype) are to prefigure (and thus eventually cause) actual *genetic* changes, unless the effect occurs in the germ line (eggs and sperm), there will be no actual evolutionary effect. But there is a second, much less direct kind of way that such changes can be passed on.

An example of this indirect method can be shown in maternal behavior. Let us take a mother rat (but the process could apply to a human mother). The mother rat had a poor upbringing. This has resulted in an epigenetically produced mark in her DNA that affects her hormones as she grows into adulthood (especially her stress hormones) and causes her, after pregnancy and birth, to also be a bad mother.[8] She does not groom or otherwise love her little pups (as baby rats are technically called). She got this epigenetic change because her mother was a bad mother. But bad is bad, and bad changed her DNA by epigenetic changes. After she gives birth, her pups are not well taken care of. The effect on them is the same as the significant environmental change that bad mothering caused in their mother's own childhood. Because of this, their own behavior when they are parents is now changed. Their own levels of various stress and other hormones are affected in ways that repeat a cycle resulting in them being "born on the bad side." *They* become bad mothers. And thus, this passes forward, generation by generation. Not by a direct transmission in marking their gametes but by their very mother's behavior, itself regulated by skewed, epigenetically caused hormonal levels.

To think that there are not direct applications to understanding humanity and the effects of poor parenting is ludicrous.[9]

The acquisition of stress in organisms causes changes other than epigenetic ones. New studies show that increased stress levels can lead to increased rates of mutation. The effect this might have on evolutionary change is unknown, but because most mutations are deleterious or even lethal, increased mutation rate in a higher-stress environment cannot be viewed as a means toward greater fitness in most cases.[10]

There remains a great deal of dispute as to the relative importance of epigenetics, the extent to which it is heritable, and even if there is anything actually novel compared to the classical, establishment view that random mutation is the prime fuel of evolutionary change and to the resulting record that is the history of life. Much of this discourse comes from the ongoing belief that what is called "reprogramming" makes the epigenetic additions of methyl molecules attached to DNA a nonfactor—in that they are erased at fertilization. It has long been "truth" that the epigenome (the complement of chemicals that modify the expression and function of the organism's genes, such as the methyl molecules that can glom onto specific genes during the life of the organism due to some environmental change) of the parent is reprogrammed (all epigenetic traces removed) twice: once during the formation of the gamete itself (the unfertilized egg, or a sperm waiting around to fertilize an egg) and secondly at conception. Erase and erase again. But now experiments definitively show that some of the chemicals added during the life of an organism do leave information in such a way that the offspring has their genes quickly modified in the same way that the parents did. The same places on the long DNA molecules of the newly born (or even the "not-yet" born) get the same epigenetic add-ons that one or both of the parents had. This is not supposed to happen. The revolution is the realization that it does. Lamarckian. Not Darwinian.

Also controversial is what Lamarck really understood compared to what later writers have inferred from his writings.[11] Lamarck seemingly intuitively understood the scientific attributes, yet they appear to be misinterpreted both from the translation of French to English and from the far different terminology from more than two centuries ago that Lamarck necessarily used. His comprehension was of his time, and thus includes ideas and conclusions now laughable, such as the impossibility of extinction, among others. But a reality is that the understandings coming from epigenetics, and especially the mechanisms of heritable epigenetics, can and now must be included under the bigger tent, or in the paradigm of evolution and evolutionary change through time in organisms.

All organisms go through epigenetic changes during their lives. Not all of these changes get passed on to the next generation. But their sum has its own term. The term *epigenome* is used to describe the composition of an original genome (the DNA with its coded genes that an organism received a first time at fertilization of egg by sperm) that has, over the organism's lifetime, acquired methylated sites or histone modification or even the transcript errors associated with small RNAs. Because the marks for these changes are progressively added over the organism's lifetime, the epigenome thus changes, but the DNA code does not. The epigenome is thus the original genetic code with markings added by events in life. Some of the marks are passed on to the next generation or even generations. Those that are passed on are called "heritable epigenetic" changes.

This difference leads to a great deal of rhetorical misunderstanding. The field called epigenetics includes both kinds of processes: the "epigenetic" changes during life and the "heritable epigenetic" changes that move through time into subsequent generations.

Many historians of science are scrambling to understand and provide historical context for this now-burgeoning and still in many corners somewhat ill-smelling topic. Historians of science see three "epochs" of evolutionary theory from Darwin onward. The first, the original Darwinian stage, established the principle of change through natural selection. The second, the modern synthesis, added the true nature of effects of heredity and also documented how actual change in DNA can be caused by recombination and mutation. We are entering an epoch in which epigenetics is added to the mix. The environment can affect how, when, and even *if* genes are expressed in both space (within the body) and time (during growth and later life) without altering the original DNA sequence.

Until the discovery of DNA, there was an understanding that inheritance came through physical units called genes, but the actual makeup of a gene was at best poorly understood. Now we have what one of the discoverers of DNA, Francis Crick, rather immodestly (but probably correctly) called the "first principle of biology": that "DNA makes RNA, and RNA makes protein."[12]

This "central dogma" seems to say that DNA is the sole source of information on what to build. Decades later, a series of control genes, such as *Hox* genes in an animal, would be found which determine what to build and when to build it. The builders (or even what has been built, which sounds strange in this analogy but was deemed important by Crick in terms of life) cannot change the blueprint for the whole structure.

We are seeing biology where the "central dogma" remains predominant but is no longer the only dogma. In evolutionary terms, the script can be changed by outside forces. Thus, "genetic control" is no longer the only determining factor of what, where, and how a protein or some life activity determined by information on DNA is being built. Environment can change things, and not only the scene in a given movie. It can also change all the sequels of that movie as well, by being heritable. No wonder that some claim that heritable epigenetics is causing a revolution in science.

TWENTIETH-CENTURY EVENTS,
TWENTY-FIRST-CENTURY CONSEQUENCES
AND DISCOVERIES

The 2012 landmark book *The Epigenetics Revolution* by Nessa Carey[13] wonderfully summarized the basic processes of epigenetics from a chemical and biological point of view, and showed the importance of epigenetic processes in current life. I propose that many of the most important events in deep time may also have been significantly or even mainly caused through epigenetics rather than through classical Darwinian evolution, as demanded by the new synthesis. These include the first formation of life; the subsequent unification of all Earth life with the same set of amino acids and DNA code; the diversification of life and evolution of multicellular life through processes of symbiotic capture of various kinds of life by more dominant, larger forms; the many rapid and dramatic formations of various and highly different body plans in the Cambrian explosion; and the recoveries and evolution of new body plans of life following the great mass extinctions.

To almost the same extent, the following case studies explore how major events in human history (as much cultural as biological) may have opened torrents of both biological and cultural evolutionary change through epigenetic pathways, and especially behavioral evolution in humanity through the evolutionary changes in levels of cortisol and serotonin, the effects of hunger, and the *MAOA* gene (aka the "warrior genes"). Nessa Carey talked about the Dutch winter: how the starving Dutch in 1945 were not only personally changed by the hideous food privation caused by the Nazis, but also how the sons and daughters of these Dutch victims themselves inherited genes that caused them to suffer from two kinds of eating disorders, either producing starving waifs or morbidly obese individuals.[14]

But if the Dutch winter starvation led to such changes into the next generation, and even generations after that, what of the other monumentally destructive events in human history? There have been many famines, such as the Irish Potato Famine and the great starvations of Biafra in the near past. And moving into other areas that can produce epigenetic changes: How did the Black Death change humanity? Survivors, whether they were infected and survived or simply lived through the hideous times of death, might have themselves undergone epigenetic change, either from the ravages the disease imposed on their bodies or from watching loved ones and friends dying in such hideous agony. Such grief changes us. And our descendants, apparently.

And what of the Nazis? How could such evil have been spawned where literally millions of men and women carried out murder, hideous murder, on and off the battlefield? The Allies, who in most recent histories are depicted as peace-loving democrats forced into learning to kill and kill again, opposed them. The Axis and Allied soldiers were all products of the Great Depression. All had been touched physically or mentally by what was the largest global change in human resource distribution probably of the last millennium.

In recent years, we Americans were revolted at the loss of nearly 4,500 of our soldiers in Iraq. Yet, in the Vietnam War, we lost more than 58,000 soldiers (some my friends) with what seems to be less anguish as a country. And by World War I and World War II standards, the entire loss of American lives in the Vietnam War was not even close to the loss of life in single battles on the Eastern Front, or in the trenches of the great "offensives" of the World War I Western Front. Perhaps those surviving soldiers were already affected by precursor events, but they were certainly affected by the wars themselves. Today, there are many names for the effects of great violence. *Post-traumatic stress disorder* is one; in the past it was called "shell shock." Only now are we seeing as the tip of the iceberg the research and far-reaching implications of how violence, war, and famine might combine to epigenetically change the biology of a person to the core of the most basic information that makes us: our DNA.

In their important 2009 essay "Is Evolution Darwinian or/and Lamarckian?"[15] Eugene Koonin and Yuri Wolf pointed out that one reason Lamarckian mechanisms were shunned by those building and guarding the evolutionary new synthesis during the twentieth century was that no one could figure out how adaptive phenotypic characters acquired during the lifetime of an organism could be, as the authors put it, "reverse engineered" back into the genome. Late in Darwin's life, scientists began a series of experiments to explicitly test Lamarckian hypotheses. The most notorious was by a German biologist named August Weismann,[16] who cut off the tails of a bunch of rats and then touted that the next generation of rats had tails, thus refuting Lamarckism. Even though it totally missed the mark (surely there is no reason a rat with no tail would be an "improvement" for dealing with its environment), this experiment was popularized and was influential in a further condescension toward Lamarck's work among the public as well as among scientists.

For Lamarck's reputation (and theories invoking his name), worse was to come in the twentieth century. Scientists seized on the most fallible of Lamarck's many predictions: that it was a drive toward progress, and ultimately perfection, that drove evolutionary change. Attempting to "prove" that the drive for progress was real, an early twentieth-century biologist named Paul Kammerer[17] tried to show that amphibians changed their color patterns based on the temperature of the water they bred in, and that these changes were heritable. He was caught having used ink to tattoo the result he wished. Falling even further, the Russian charlatan Trofim Lysenko[18] picked up on Lamarckism as a justification for communism and was given state support. One of his experimental results claimed that feeding a cow chocolate and butter would cause it and its descendants to produce fat-rich milk. Russian scientists who attempted to laugh these results out of science paid with their lives, contributing to the wholesale purge of an entire generation of Russian biological science.

In this century, many of Lamarck's conclusions can now be demonstrated by testable science. Yet virtually all of these concern physical changes; few to date have dealt with genetically influenced behavior. This is the still-great unknown for humanity. How much can human behavior, the good and the bad, become heritable from environmental effects during the life of a pre-reproductive or actively reproducing human? Can war and widespread violence, violence against an individual, major disease affecting entire populations, or famine and community-wide starvation cause heritable change among survivors?

Lamarck was indeed sure that behavior was a major part of his overarching theory about evolutionary change. As noted by Koonin and Wolf,[19] Lamarck viewed heredity as a three-part causal chain: An organism encounters an environment, that environment causes behavioral change, and the behavioral change causes change in form. In the face of major environmental change, organisms respond first by changing habits. The change of habits then produces change in morphology. Lamarck wrote: "Whatever the environment may do, it does not work any direct modification whatever in the shape and organization of animals. But great alterations in the environment of animals lead to great alterations in their needs, and these alterations in their needs necessarily lead to others in their activities. Now if the new needs become permanent, the animals then adopt new habits that last as long as the needs that evoked them."[20]

The discovery of DNA was one of the fundamental discoveries in the history of science. DNA is made up of a series of instructions. Each of these instructions is used to build things. While it was originally thought that genes were simple on-off switches, now we know that there are complex controls on not only what is built (for instance, how to build a hemoglobin molecule for our blood) but when to do it and how much of it to build.

In a recent article about this new field, Mark Rothstein, Yu Cai, and Gary Merchant provided a useful analogy: "The genetic code has been compared to the hardware of a computer, whereas epigenetic information has been compared to computer software that controls the operation of the hardware. Further, the factors that affect the epigenetic information may be analogized as parameters for operating the software."[21]

As noted earlier, the epigenetic mechanisms can come from events affecting the organism as a whole *during its life*. This is a Lamarckian kind of change.

What should now be far more rigorously tested is not just the possibility that the characteristics acquired and made heritable are manifested in the change in morphology or the size of an organ or a particular body shape, but that various kinds of *behaviors* themselves become heritable through epigenetic changes that are then subject to natural selection.

CHANGING THE PARADIGM

Major breakthroughs in the field of genetics and DNA technologies, such as gene splicing, have revived interest in the Lamarckian paradigm because many scientific results can no longer be explained by Darwinian theory alone. One of the most important of these is the discovery that sometimes great swaths of DNA can be rapidly inserted into another organism, utterly changing almost every aspect of its biology. These changes are Lamarckian, not Darwinian, and they are important mechanisms affecting the history of life. The most important of these is called lateral gene transfer (LGT; sometimes called horizontal gene transfer). This occurs when entire genes and even suites of genes along hunks of foreign DNA are inserted into an organism's genome by biotic invaders.

But it is not only the successes of LGT but the fact that this mechanism was so successful that caused early life to invent a defense against it. That defense system,[22] which appeared in deep time among the oldest lineages of single-celled microbes,[23] is also Lamarckian. And, as we will see, its elucidation has unexpectedly given geneticists the single most powerful weapon in their arsenal for changing the DNA of an organism during its life, or, more powerfully, before its birth.[24] It will also radically change the nature and future of life on Earth.

The defense system is known as CRISPR-Cas.[25] To defend against the insertion of a foreign length of DNA, with the actual invader being perhaps a virus or prion or another bacterium, the invaded host uses a different, foreign hunk of DNA, placed at a specific site, and a product built from this new gene to specifically hunt down and destroy the foreign strand of DNA. In so doing, the successfully defending cell has changed its genome in ways of its choosing, not in the ways that the invader wished to happen. Thus, the defense against Lamarckian change is to produce a different kind of Lamarckian change.

But here's the irony: Humans have figured out how to use the "find-and-eradicate" method in a way that allows biologists to find and destroy, or find and replace, genes with a set of new genes of their choosing. Genes that might make a human no longer affected by life-shortening genetic diseases. Or genes that prevent mushrooms and other fruits and vegetables not from bruising while being transported from field to store, and later rotting on supermarket shelves. Or genes that keep the muscles of dogs in proportion to the rest of their bodies. The technique has been given the name CRISPR-Cas9. It is already revolutionizing biology and is hailed as the tool to make some of the most significant and life-enhancing procedures in the field. A "godsend" is how many describe CRISPR-Cas9, in language not far removed from the praise that followed the early publications trumpeting the new era unfolding before humankind by the discovery of nuclear fission—manna not from heaven but from the work of nuclear physicists. Unleashing the atom! Unleashing the ability to put new genes in place in human embryos! What could go wrong?

The use of this system as a tool is just beginning. We will come back to it near the end of this book in a discussion about future human evolution. Its importance was recently described in the conclusion to a summary of the method, and is useful here, simply because this process is absolutely Lamarckian: "The rapid progress in developing [CRISPR-] Cas9 into a set of tools for cell and molecular biology research has been remarkable, likely due to the simplicity, high efficiency and versatility of the system. Of the designer nuclease systems currently available for precision genome engineering, the CRISPR/Cas system is by far the most user friendly. It is now also clear that Cas9's potential reaches beyond DNA cleavage, and its usefulness . . . will likely only be limited by our imagination."[26]

Yet the fear among many is that this new and simple process can be dangerous both by intent (biological weaponry) and through incompetence (unregulated use leading to genetic "accidents" every bit as dangerous as the meltdown of a nuclear power plant). Lamarck's revenge indeed, if his ghost is embittered against the humanity that so tortured him in the final blind, poor, and starving years of his life.

THE ROLE OF ENVIRONMENT

Real estate success (and failure!) is ruled by three words: Location! Location! Location! And so too, in the natural history of many species, it might be paraphrased that human health is similarly ruled by environment, environment, environment! The old nature-versus-nurture argument is taking a turn toward nurture.

Take a pair of identical twins who have undergone quite different biological and perhaps behavioral lives (since it is increasingly clear that some human behaviors are heritable). Put one of the identical twins in the best penthouse condo available, with experience of food, drink, exercise, massage, vacations, little work, and less stress—life in the bubble of wealth. Put the second twin in the kind of "life-challenging" environment that poverty always presents. Revisit each twin after thirty years. If our impoverished twin has survived the random violence and has avoided stroke or early heart attack or diabetes or any of the cancer windfalls coming from a poor diet and exposure to tainted air, lead-filled water, high levels of biochemical poisons such as PCBs and estrogen mimics, and on and on, we will find two very different kinds of creatures.

Because they are identical, these twins began life with the same genetic code. Gene for gene, they are indeed identical right down to level of molecules. And yet they will probably look strikingly different. In technical terms, we say that these physical differences are manifestations of different phenotypes arising from the same genotypes.

This happens because genes are not the ultimate dictators of phenotype. Environmental conditions can and do dictate whether a gene does its normal job (or is expressed) or if, instead, something turns it off. A specific protein that was called for is not produced, in fact. And the reverse happens too. Sometimes lethal genes sit unmolested amid all the needed genes along the DNA helices. But they are chained and held dysfunctional. But along comes an environmental trigger and they are switched, even though natural selection, good old Darwinian evolution, had long ago tamed this killer into being normally switched off.

We are not our genes. We are the products of what the environment does to our genes. There are many examples. Even more insidious things happen when cells stop doing their normal job, stop reproducing themselves with fidelity or repairing themselves, and instead go haywire because of exposure to some kind of poison, or even too much of a normally good thing, like too much oxygen. The results are runaway generalists good at nothing but fast growth—renegades overwhelming the surrounding and staid workers, like functioning liver cells or brain cells or thyroid cells, through a numbers game that finally causes the organ or tissue that they are part of to cease normal function. We call these cancer cells, and rightly they are the most feared of all biological cells, killers usually swift and painful as they shut down bodies, organ by overwhelmed organ, killing off everything except the nerves. A simple mercy would be for the pain responses of the invaded portions to at least shut down. That does not happen, as we all know too well, for who among us has not been either infected by or touched by cancer in a friend or loved one?

The analogy of the twins—one in a rich, healthy place and the other in a polluted, environmentally poisoned place—is in many ways the story of our own species evolving as small bands on the African veldt and then setting off on an Earth-wide walkabout beginning two hundred thousand years ago. Life was probably short for many reasons, not least of which were the presence of human-eating predators, interband warfare, disease, and especially bacterial infection in a time long before bandages and antibiotics, when any major wound could easily end up becoming infected to death. Yet these earliest *Homo sapiens sapiens* (or the "Moderns," as some call us) lived in a world without atmospheric gases that are carcinogenic or waters filled with toxic chemicals or food that is "processed" with salts, nitrates, and preservatives, which are nothing less than analogues of the chemicals we use to keep bodies from rotting in medical schools around the planet.

Our present world, and more specifically human civilization, is awash with more epigenetic change–causing environmental agents (known, such as toxins, and inferred, including the many various kinds of environmental stress molecule formations believed to cause heritable epigenetic behavioral change) than perhaps in all of the past 200,000 years of our history combined. This is certainly in the effects of toxins, as well as those of smoking, "illegal" drugs, and perhaps the rampant use of computers and cell phones. We are in a warmer atmosphere than at any time since the cognitive revolution of 70,000 years ago, and perhaps warmer even than the several-century-long warming more than 100,000 years ago, an event that produced a short-term sea level rise causing the melting of the Antarctic and Greenland ice sheets to the tune of a three-meter sea level rise. We are surrounded by the chemicals produced to keep 7 billion people in cars, in houses, warm, on telephones, killing other humans, and growing and eating the food necessary to feed most of us. The future evolution of humans is not in the future. It is now. It is to some extent, perhaps to the largest extent, epigenetic.

Summarizing Epigenetic Processes

There are several categories of processes involved in epigenetics. We can start with those that have been called "DNA modification."

1. *Methylation*—Lengths of DNA can be deactivated by attaching small organic chemicals called methyl groups, which can inhibit the production of proteins called enzymes that are used to build other proteins (often by making possible or speeding up chemical reactions) necessary for life functions. In essence, these methyl groups are on-off switches that were not present before. They bind in specific places on DNA, where cytosine is adjacent to guanine; cytosine and guanine are two of four chemicals used in the DNA code. They are the "rungs" of the DNA ladder.

2. *Modifications of gene expression*—There are many ways that DNA can be modified to cause epigenetic change. In essence, virtually anything that modifies gene expression, such as increasing the rate of protein production called for by the code or slowing it down or even turning it on or off.

One of the most important of these on-off switches in mammals is known as "X chromosome inactivation." Because the females of mammals have two X chromosomes compared to only a single X chromosome in mammalian males, females can bring more genes forward in reproduction, depending on which of the X chromosomes of the female is used in the fertilized embryo. Without some sort of regulation, the female adds a greater "dosage" of the genes on the X chromosome. This can be controlled by using methylation to turn off the genes from the extra X chromosome, which is an important epigenetic effect. Epigenetic effects change the fate of individual cells during the development of the fetus from a single, fertilized egg into a vast assemblage of cells of many different kinds. This happens in three or four stages, to the further, mostly unchangeable type of cell needed. But cells are also responsive to major environmental stress during life, and some genes in the cells of an animal or plant ultimately change gene expression during the lifetime of the organism in response to the environment, and these can also become heritable changes.

3. *Reprogramming*—Geneticists use the word *reprogramming* when talking about how the epigenetic marks accumulated by a future, reproductively mature male or female are *erased* prior to being copied onto the genetic code when they eventually produce offspring. This has long been assumed to always be the case. Yes, it was discovered that (especially) methylated DNA accumulated during the life of an organism (including humans), but dogma was that when sperm and egg are formed, and when fertilization takes

(*continued*)

place, the slate of these Lamarckian additions is wiped clean not once but twice: that the epigenome (genome that has acquired methylated sites or histone modification or even the transcript errors associated with small RNAs) changes back to the original DNA clean of the chemical freeloaders. The epigenome is thus the original genetic code with markings added by events in life. The parent is reprogrammed once during the formation of the gamete itself (the unfertilized egg or a sperm waiting around to fertilize an egg) and again at conception. Erase and erase again. But an increasing number of studies demonstrate that this is not always the case. Reprogramming, or erasing, is thus not as thorough as once thought.

EPIGENETICS AND THE HISTORY OF LIFE

A surprising conclusion when reading the increasing library about epigenetics is how little effort has been made to use its evolutionary implications for larger-scale questions about the history of life. Even more surprising is how there seems to have been little or no interest in taking these implications and asking about trends in human history.

We have good knowledge now that major events in any human life, such as starvation or major violence or great emotional trauma or religious conversion, can cause epigenetic changes, and that some of them are heritable. Yet there has been no attempt by biologists to make even conservative estimates about what the great plagues of humanity—the wars, plagues, and famines—might have done to successive human generations beyond mere population subtraction. So too have the great increases and decreases in organismal diversity been ignored in any Lamarckian context.

It *has* been recognized that epigenetic mechanisms, whereby gene activity is regulated without altering the DNA sequence, can result in heritable changes in an organism. The greater the environmental change, the greater the chance of epigenetic as well as evolutionary change, or so think some of the most experienced biologists who have expertise in both Darwinian and some flavor of Lamarckian modes of evolutionary change. Yet applications of this understanding to interpreting the history of life, as well as the history of humanity, are few. But of these, Eva Jablonka and her various co-authors[27] have been making substantial contributions. They champion the hypothesis that epigenetic inheritance would be *favored* in what might be regarded as the comparative contributions of two separate evolutionary processes: epigenetic and Darwinian forces producing evolutionary change. The epigenetic processes might be especially important in the evolution of microbes.[28]

While we animals believe we run things, it is the microbial communities that dictate almost all aspects of the geobiological effects on atmosphere and ocean chemistry. Microbes produced the oxygenated atmosphere some 2.5 billion years ago; microbes almost eradicated animal life in four separate mass extinctions in the past 500 million years. And if the microbes are feeling a bit of hubris, in all probability their world is run not by them but by the hundreds of viruses that each bacterium often carries either on the outside or inside.

As will be noted in a later chapter, the origin of life itself probably involved epigenetic-like processes. There remains a series of questions of when and why the specific epigenetic processes, such as methylation, first appeared. Of the three most common epigenetic processes—methylation, histone modification, and the evolution of RNA interference by small RNA molecules (RNAi)—the oldest may have been RNA life. The RNAi systems may have evolved as a response to a variety of parasites (such as viruses) that tried to mine the nucleic acids, the earliest vampires. Now the RNAi effects appear to be more pronounced in eukaryotes, from tiny yeasts to large animals that have the large cells containing a nucleus and other intracellular organelles characteristic of this major domain of life, Eukaryota (the others being Archaea and Bacteria).

RNAi perhaps have a far greater role in evolution than has been accepted by all but the truest believers of the importance of heritable epigenetics.[29] These tiny molecules may be crucial in the conformation of chromatin. It is this DNA shape (as produced by different shapes of chromatin), after all, that along with methylation is known to be a driver of epigenetic change. They may also target specific rungs on a DNA ladder, changing them so that when that particular DNA molecule replicates, the change is also replicated.

From RNA life to DNA life, from single cells to multicellular, from algae to animals: the history of life is a vast panorama. It is virgin territory to be examined through the lens of epigenetics. One of the most exciting vistas concerns the times immediately after mass extinctions, when new kinds of animals rapidly repopulated Earth.

We cannot *know* that the rapid evolution of what is called the "recovery fauna" in the times immediately after the great mass extinctions of the geological past were because of the appearance of epigenetic mechanisms producing new organisms at a pace faster than is or was possible through Darwinian evolution. That inference will be made when trying to understand how the early Paleocene-age mammals in the first five million years following the total extinction of nonavian dinosaurs were able to evolve into so many kinds and body plans so quickly. This and other examples of rapid evolution following mass extinction lead to the hypothesis that such examples were importantly influenced by contribution of epigenetic mechanisms.

The argument, at least, might be novel: that in deep time, geologic time, there were two kinds of evolutionary change—"gears," to use an automobile analogy. During "normal" times, those long intervals making up most of geologic time, we have the slow and steady drumbeat of random mutation, the gears of the so-called molecular clock used so widely in estimating the times of divergence of lineages in deep time. (By comparing the DNA of somewhat similar animals, and assuming a constant rate of mutation, the time of divergence of the two species can be, and routinely is, estimated.)

But there were *extraordinary* times, in the true sense of that word. Times when the environment went atypical, and atypical too was the rate at which new species formed in response. Times such as a sudden-plunge into a global ice age (over a few thousand years). Or immediately after a "greenhouse extinction," the consequence of rapid global warming caused by volcanically produced greenhouse gases heating the planet to the point that the oceans lose their oxygen, thus killing the preponderance of species. Or an asteroid hitting the Earth, as happened 65 million years ago.

After those events, the fossil record tells us that not only new species but entirely new kinds of body plans repopulated Earth. It is the revolution of the now far-cheaper and far-more-accurate geological dating available in this new century that gives us quantitative proof of how quickly these post-extinction biotas came into existence. Too fast for Darwinian evolution, but possible by *Lamarckian* evolution. By heritable epigenetics, to be more accurate in terminology.

This is revolutionary. The history of life should be subdivided into times of evolution dominated by Darwinian mechanisms (most of the Phanerozoic—the 540 million years of common fossils) and far shorter intervals dominated by epigenetic mechanisms: the Cambrian explosion; the Carboniferous explosion of the one, high peak of oxygen in Earth's history, some 300 million years ago; after the Big Five mass extinctions; and during intervals of what is known as the "true polar wander," when the entire globe moved at extraordinary speed, taking arctic environments into the warm, and the tropics into the cold. It is also the ticket for paleontology to come back to the high table of evolutionary theory.

EPIGENETICS AND HORMONES

We can conclude this chapter by looking at what appears to be the common link in both the history of life and the history of civilization as influenced by epigenetics: the role of epigenetic processes in the evolution and influence of hormones, and, most importantly, stress hormones. This subject remains a point of contention among the critics of epigenetics.

An unfortunate aspect where the history of science raises its ugly head is that the first scientist to link a role of stress hormones in epigenetics was Paul Kammerer, whose later faking of data is used to try to discredit all of his work. Kammerer exposed rats to high temperatures to see whether their offspring were affected through increases in both morphological and physiological variability that were not seen in those rats whose parents were not so exposed (the control group).[30] Heat is one of the most potent initiators of stress molecule formation. The study of specific "heat shock" hormones is a major research area in this time of global warming, especially in fish. But since Kammerer's study, evidence makes that link. When stress hormones are produced by sudden environmental changes of important intensity, rapid evolution occurs quite often. When the epigenome is changed, so too is the evolutionary "trajectory" of the organism experiencing the profound environmental changes during its life.

Among the many experiments showing this, perhaps none is more significant for understanding the near future than the long-running breeding experiments in Russia on silver foxes.[31] These foxes were bred and selected for tameness. Foxes are not stupid. There must be great stress in seeing five or more of your siblings suddenly disappear, generation in and generation out. But the variable hormone levels of the pleasure molecule, serotonin, affected the gene that is associated with aggression. The great surprise was that while hormones were affecting a specific gene, other biological effects were triggered, the most obvious of which were the genes affecting fur color. In some foxes the result was a novel coat with white spots.

For humanity, there is another study that portends something much darker than color change. Another kind of stress is the exposure to toxins. In one study, rats exposed to known carcinogenic toxins produced a large variety of genes and DNA sequences that became methylated, and these epigenetic changes were passed on to several subsequent generations.[32]

THE SCIENTIFIC PRESENT

The biggest controversy about epigenetics, or at least the part most important to epigenetic inheritance, is whether behavior itself can be inherited, especially when that behavior is a product of trauma in a lifetime. In humans, for instance, what is the possibility of epigenetic inheritance of post-traumatic stress disorder? While refuted by the various global militaries for decades, this acquired mental state provokes real physical changes in the body.

Provoking "fear" has its own number of rat-torturing studies. One of the most eye-opening was a study in which mice were taught to fear a smell of substances that should not elicit fear (such as toxins that have a smell). A certain smell that otherwise is neither a positive or negative stimulus is associated with fear of pain. In this case, the aroma from cherry blossoms was associated with shocking the poor mice's feet with electricity. The astounding discovery was that fear was inherited by the next generation of mice. It was this kind of study which more than anything caused evolutionists to reconsider whether the current theory of evolution needed to be added to in order to accommodate the neo-Lamarckian findings.

CHAPTER V

The Best of Times, the Worst of Times—in Deep Time

T HERE is a conception that what is often called the "history of life" is pretty well mapped out in terms of the when, what, and where. But almost completely unknown remains the "why." And, from what science seems to be telling us in ever more interesting fashion, the "how" of this history, the actual evolutionary mechanisms, might be poorly known in many important details as well.

Since animals made their appearance in the fossil record in any abundance (in the Cambrian period), they are well mapped out in terms of the "higher" taxa: the appearance of the phyla, the classes, the orders, and most of the families. But there remain very interesting discoveries still to be made about many transitional species yet unknown. The discovery of the "fishopod" named *Tiktaalik* of more than 300 million years ago was such a discovery (more than a decade ago), a true "missing link" between fish and amphibians.[1] A similar gap was filled in with the discovery of fossils discovered in Pakistan strata that dated to more than 50 million years in age. These fossils appear to have been true "missing links" between land animals and true whales. In many such transitions, however, from one kind of body type to another that is quite different, we have neither the transitions forms in the fossil record nor a true grasp on why these evolutionary changes took place.

Among the most radical of environmental changes affecting various groups of life were the many challenges faced by organisms making the jump from a wholly aquatic existence to one mainly or completely on land. This is true of both plants and animals. With the colonization of land by plants predating the first animals (with plants occurring on land perhaps 500 million years ago or more), and finally with the first trees and forests some 400 to 360 million years ago, land offered a new habitat with new resources. There was plant life for herbivores and abundant invertebrate animals, including hordes of insects, to attract carnivores. The current supposition is that the first land vertebrates, amphibians by definition, came about through known processes of natural selection—by Darwinian evolution, in other words. This supposition should be reexamined in light of the new research into the rapidity of morphological change that can be caused by dominantly epigenetic, instead of Darwinian, evolution.

ENVIRONMENT AND THE HISTORY OF LIFE

The role of the epigenome and its ability to rapidly affect the results of genes in times of great environmental change must be seen as one way that life has responded to the vast environmental changes of our evolving planet over the last 3.5 billion years.

Planets are dangerous places even in the best of times, and much more so in nastier times. When a large (let's say five miles in diameter) asteroid hits a planet with a velocity of fifteen miles a second, life near and far will take notice, and mostly die. One of the great passages in Tom Wolfe's book *The Right Stuff* is a description of American military pilots involved in testing the giant jet fighters being built in the 1950s. The mortality rate among those pilots was staggering. Wolfe describes a theoretical situation that must have happened over and over: the plane, from some kind of mechanical failure, goes into a steep and rapid dive. As Wolfe describes it, the pilot, facing imminent death, calmly "tries process A, tries B, tries C, tries . . ." and then boom. The point is that when the pilot is facing "environmental crisis" (i.e., the plane falling to Earth at hundreds of miles per hour), the more options the better. But also key to that passage is the *calmly*. Life was born from environmental chaos. It would be strange if life has not encoded a way to rapidly try A, B, C, D, etc., when faced with imminent death by extinction, the anthropomorphized *calmly* being the epigenetic mechanisms trotted out amid environmental crisis.

Now put this in the context of Earth life some 65 million years ago immediately after the impact of the asteroid that hit Earth in what is now the Yucatán region of Mexico. Let's mix the metaphor a bit. This is not about the "options" that the dinosaurs and so much else had in the minutes to days after the impact but about the options that various kinds of life had in the weeks to months to years after. The world utterly changed. Global darkness. Plummeting temperatures, rotting meat and disease and bad water and no photosynthesis. Here is where the epigenome kicks in. The production of lots and lots of different kinds of creatures from the same set of genomes. It is not "Try A! Try B! Try C!" It is "Try them all! See which one works!" This could be different shape, physiology, size, behavior, whatever. The point is to take as *many* kinds of phenotypes as possible and worry about the genes later.

Life can and does respond to great environmental change by having mechanisms to vastly increase phenotypic possibilities without a necessary, corresponding increase in genotype: This is the great new understanding coming from the study of epigenetics, and why evolutionary theory, and especially the interpretation of past times and past catastrophes, requires a theoretical updating. Changing genotype sufficiently is hard to do in the short timescales that environmental crises require. But the epigenome can produce a wide variety of new shapes, sizes, morphologies, and behaviors during significant environmental change. In such times, producing new phenotypes is paramount; natural selection can sort it out later.

It is crucial to interpret the evolutionary record of complex life soon after global environmental crises. At such times, great plasticity (in morphology or other traits, perhaps most crucially in the factors that lead to various kinds of behavior) not only allows organisms to cope in new and really nasty environmental conditions but allows them to *generate* traits that make them more adaptable. When the world has gone dark for six months, when the global temperature has dropped from lushly warm to high-latitude darkly cold, mechanisms that produce as many different kinds of traits as possible would be hugely useful. Through heritable epigenetics, there can be a wholesale flush of new varieties of organisms. But the difference to traditional evolutionary theory, and Darwinian evolutionary mechanisms, is that *the trait comes first*; genes that cement it follow, multiple generations afterward.

Life—but also, obviously, human history—seems to follow a well-worn dictum long noted about military life during wars: "Long periods of boredom punctuated by short intervals of terror." So too in the deep-time history of life on earth. There were long periods when not much happened. Then peace (periods of sluggish evolution) was shattered. The times of environmental change and mass death first caused high extinction rates, followed by heightened rates of new species formation. Thus, the perhaps novel intuition posed here: that deep-time evolutionary history perhaps should be divided into the calm, slow changes brought about by Darwinian mechanisms of chance mutations gradually accumulating versus shorter times of widespread environmental chaos, during which it is the production of vast and vastly different kinds of phenotypes that prevails.

DARWINIAN VS. EPIGENETIC TIMES

Paleontologists are quite secure in asserting that mass death opened the door for new species, and quite often new kinds of creatures with radically different body plans. The wave of small, rat-sized, and rat-shaped mammals taking over amid the rotting carcasses in the Paleocene epoch could not look less like dinosaurs. But while paleontologists are sure about *why* a wave of post-extinction "recovery fauna" mammals and birds appeared, they remain flummoxed about *how*—at least in terms of evolution. The clear answer is the epigenome. At this time, epigenetic mechanisms flooded the world with a hugely diverse assemblage of small rat shapes, teeth, as well as a wide range of new and varied behaviors in feeding, defense, reproduction, territory acquisition, social structure, and on and on. In the centuries after the Chicxulub Impact, all these new attributes and behaviors became a part of an even *larger* assemblage of phenotypic characters.

Environments obviously change today, and change in different ways and at different rates depending on the environment. Currents shift, mountain ranges rise or fall, and climate changes accordingly. The size of the ocean basins enlarges or decreases depending on the heat flow coming from within the Earth and acting on the seafloor and spreading ridges. Again, these are very slow changes.

Yet some intervals of time not only had slow changes but had superimposed upon them very rapid global changes that would have affected life. The recently concluded ice ages (or at least concluded so long as there is an industrial civilization on the planet) were accompanied by a succession of radically changing environments, with the changes occurring at decadal time periods. The speed of the melting ice some 14,000 to 12,000 years ago was accompanied by fast changes in climate as well as a global average temperature increase and a rapidity of the rise of the sea, with a rise of over one hundred meters in perhaps two millennia: These were times of extraordinary environmental change. Yet even a few millennia before the last of these periods of glacial advances and retreats that began about 2.5 million years ago, say from 24,000 to about 18,000 years ago, amid the frozen world, change would have been suppressed compared to before and after. It was a time of stability, if a very cold stability.

Going back to the time immediately prior to the start of the Pleistocene—the Ice Age, as it is often now called—there was a far longer time of stability. For tens of millions of years, land environments were highly stable, with little change in global temperature beyond a slow decrease that was occurring in hundred-thousand- to million-year increments; there was virtually no change in global atmosphere oxygen values, only a slow change in the positions of the continents, and even stability in sea level. This was a time of extraordinary continuity in environments. Global carbon dioxide levels were more than four hundred parts per million. There was no arctic sea ice. The ice sheets that were present were far less voluminous than now, and sea level was far higher because global temperatures were far higher. Average summer temperatures in the Arctic Circle, for instance, were between 10 and 15 degrees Fahrenheit higher than now. An irony is that in 2017, the driver of that warmer world, the amount of carbon dioxide in the atmosphere, for the first time in 3.5 million years was once again over four hundred parts per million. This fact alone will drive more evolutionary change, including the influences of heritable epigenetics, than any other aspect of the environment. It will also, in all probability, be the single greatest driver of future human history. But that will be determined.

One goal of paleobiology, as well as evolutionary biology, ought to be far better communication with those studying past environments on Earth from the physical science point of view. Paleobiology and evolutionary biology are usually done by different scientists, trained in different methods, and working in different departments that themselves are usually stuck in different buildings on university campuses. The two do not even use the same scientific language in many cases; the former uses the language of morphology and geochemistry, the latter the language needed to describe genes and DNA. Paleobiologists primarily use the fossil record, whereas evolutionary biologists use genetic analyses at the molecular level as their go-to tool. Both are needed if many of the unanswered questions are to be answered.

With an essentially neo-Lamarckian set of theories now accepted by many evolutionists, the questions should now be about the relative importance of the two different processes: The Darwinian explanation is that there are five major forces that cause evolution: mutation, "genetic drift," gene flow, genetic recombination, and natural selection. With a new Lamarckism added, the questions now should be about the relative importance of the "Darwinian Five" relative to and heritable epigenetics in the history of life.

The most interesting implication, and then scientific question, is whether Darwinian or Lamarckian modes of evolution change in frequency depending upon time and environmental circumstance over geologic time intervals. Because major morphological change can occur markedly faster by epigenetic compared to Darwinian processes, the logical conclusion is that the long, slow changes allowed by gene flow and natural selection working on the phenotypes of lineages undergoing genetic change by random mutations and mistakes in chromosome copying during reproduction predominate during the environmental "good times" but are inadequate in the face of either more rapid or more extreme environmental changes during our planet's "bad times" (at least as far as life is concerned). A landscape being carved out by a river cannot make any sort of response so as not to become a canyon; a rock type being heated by a rapid increase in geothermal gradient beneath it has no mechanism to avoid becoming metamorphosed into an entirely new kind of mineral. But life can adapt.

It has been proposed that evolutionary change effected by epigenetic mechanisms can be more than an order of magnitude faster than Darwinian evolutionary change in a population meeting the same environmental challenges. Some say this can be three orders of magnitude, or up to a thousand times faster.[2]

Faster in terms of morphological, physiological, or ontogenetic (the growth to adulthood) change toward organisms better suited to survive a new world. A new environmental world, such as one where oxygen was lowering or, more commonly (yet related), where temperatures were becoming radically warmer through an increase in atmospheric greenhouse gases such as carbon dioxide (and thus, soon after, by an increase in oceanic acidity, because the uptake of ever more carbon dioxide by the oceans increases their overall acidity, much to the distress of organisms that make calcium carbonate shells, such as the larval oysters now dying at microscopic sizes in our world because of the rapid runup of atmospheric carbon dioxide followed by oceanic carbon dioxide levels).

Most of the mass extinction events were caused by physical rather than biological changes. There were, potentially, many kinds of these. Among the "catastrophes" possibly initiating mass extinction were rapid and large-scale changes in oxygen and carbon dioxide levels (that were sometimes caused by life itself), changes in ocean and air currents caused by changing continent and ocean sizes and positions, increasing or decreasing rates of continental movement, the initiation of sudden episodes of flood basalt volcanism, a sudden asteroid or comet impact or comet showers, periods of intense solar activity (especially prior to the formation of Earth's ozone layer), and times of rapid reversal of Earth's magnetic field.

The thesis here is that the comparative interplay of natural selection (acting on the genome) and epigenetic selection (acting on the epigenome; or the formation of multitudes of phenotypes by epigenetic processes that are then selected on)[3] is to some degree a consequence of both rapidity as well as degree of some combinations of environmental change experienced during the life of the living species.

Good times and bad times, which really come down to rates of change.[4] Good times are times of stability. Bad times are the opposite. Change comes in the form of temperatures, global and local; atmospheric oxygen and carbon dioxide levels, which exert different effects and interact with each other through the dissolution of oxygen and carbon dioxide into water, as well as the greenhouse effects of CO_2, which affects concentration of both; the chemistry of the oceans in terms of acidity, sea level, and connectivity to smaller, shallow continental seas; and the rates of movement of continents as well as microcontinents, which then affect weather patterns. Or climate disasters such as the first and relatively sudden appearance of a monsoon where such a weather pattern had previously not happened, producing for the first time six months of steady rain in a region that had seasonal dryness, or the converse, a drought that would not break in regions accustomed to heavy seasonal rains. Such changes are the kind that might provoke rapid evolutionary change.

Based on observations over the last century, it is clear that weather changes are not necessarily slow changes. For example, the weather pattern known as El Niño is present or it is not. The monsoon comes or it does not.

So too with the other rapid environmental changes that happily have not happened in human history but that have left abundant geological and biological evidence of their sudden appearance. For example, following the asteroid impact of 65 million years ago there was one year of darkness. Even the onset of the gigantic flood basalts that are implicated in the mass extinctions at the end of the Devonian, Triassic, and Permian periods would have quickly caused major changes in weather and most importantly in local temperature and water availability. While not as fast as the effect of an asteroidal impact, the first eruption of what are called the Siberian Traps some 251 million years ago caused global change within a decade that was sufficient to begin the greatest of all mass extinctions of the past 500 million years, although earlier in time, far more catastrophic mass extinctions surely took place,[5] such as the sudden onset of the global freezes that became what are termed "snowball Earth" events happening more than 2 billion years ago, but more important biologically about 700 to 600 million years ago.

And now? The chemicals being inserted into our oceans and atmosphere at the rates we witness, added to the most rapid increase in atmospheric carbon dioxide in planetary history, are changing the trajectory of human evolution. The future evolution of life on planet Earth is being rewritten with every chemical spill, every cubic meter of Antarctic and Greenland ice melted, every species dying from habitat disruption.

It can be asked if any of the largest of such changes—such as in atmospheric gas, global temperature, and global toxins (at least to some kinds of life, such as the highly poisonous gas hydrogen sulfide)—have been associated in repeated changes to the history of life as measured by either the change in the number of species (diversity) at the global level and/or the change in the number of separate plans (disparity). Two other environmental fluctuations have changed the history of life. One was movements of Earth's entire crust at rates far exceeding the known rates caused by continental drift. This process has been termed "true polar wander" but is now more accurately known as "mantle wander." The second relates to the rapidity of reversals of Earth's magnetic field, which recently has been proposed to have a major effect on global oxygen levels, which impacts diversity.[6]

THE WORST ENVIRONMENTAL DISASTERS
FOR LIFE

The first snowball Earth episode (beginning at about 2.35 billion years ago) seems to have been caused by life: The explosive rise of cyanobacteria caused a reduction in the greenhouse effect of the atmosphere's methane and carbon dioxide content. The start of the second, and final, series of snowball Earth events began 717 million years ago and ended 635 million years ago.

Both of the differing snowball Earth episodes (each made up of ocean-freezing and then ocean-thawing events) caused a severe decline in marine organic production because the sea ice blocked out sunlight. Thus, the amount of life on Earth, as measured by its overall mass (known as biomass), shrunk to tiny values compared to both before and after the events themselves. The succession of snowball glaciations and their ultra-greenhouse terminations during both the event of 2.35 to 2.22 billion years ago and the one of 717 million to 635 million years ago must have imposed a severe environmental filter on the evolution of life.[7] The fossil record provides few clues, but the tiny one-celled organisms called acritarchs (planktonic organisms of small size that had a skeleton, and hence fossils) waxed and waned in both diversity and abundance.

Many living organisms are known to respond to environmental stress by "reorganization" of their genomes through epigenetic processes that are a direct result of the effects of the local environment on an organism, and any snowball Earth would have been stressful to say the least. The fact that diverse fossils of more complicated organisms than were there before the onset appear in the immediate aftermath of the snowball glaciations supports the notion that the snowball events created some sort of an ecological trigger for vast changes in the complexity of life and its diversity.

Glaciations themselves are suspiciously coincident with some of the most profound biological events in terms of change in diversity as well as change in "disparity," the number of new body plans, and the removal of already existing body plans during environmental crises.

Global glaciations cool the planet in a negative feedback: The more ice there is on the land and on the sea, the more sunlight is reflected back into space instead of being absorbed by land and seawater. The tropics shrink, weather goes crazy, with gigantic windstorms common along the edges of ice, and the atmosphere is filled with clay and dirt.

There have been five major glaciations that affected life: Snowball Earth 1 (2.5 billion years ago), Snowball Earth 2 (720 to 635 million years ago), the Ordovician glaciation (460 to 440 million years ago), the Carboniferous-Permian glaciation (300 to 270 million years ago), and our own, recent Pleistocene glaciation. Each has been associated with (in time) some of the most important biological innovations as well as biological catastrophes: The first snowball Earth event was coincident with the start of the Great Oxidation Event, which was the first outpouring of oxygen into the atmosphere and oceans. It was a result of the onset of photosynthetic processes in life. The second snowball earth event was also of probable biological origin, in this case the diversification of multicellular plant life; it appears to have been the trigger for the first appearance of animal life. The third glaciation, in the Ordovician period, was coincident with one of the Big Five mass extinctions. Yet it also paved the way for the first appearance of oceanic communities that appear "modern" in the nature of how food webs are constructed—in the proportion of filter feeders to grazers to predators, for instance—as well as the start of coral reefs. It reorganized the oceanic communities and set the stage for the conquest of land by plants. The Carboniferous into Permian glaciation also radically changed life, especially life on land. Prior to it, land life was composed mainly of amphibians and primitive reptiles. By its end, advanced communities of "mammal-like reptiles" (the predecessors of us mammals) had spread across every continent as had forests in almost all land environments, and reorganization in seas had taken place as well. The last glaciation, the one that has just finished relative to the age of the Earth, has also changed the planet. Prior to the onset of this Pleistocene Ice Age, beginning 2.5 million years ago, most continents had mammalian faunas looking like those in Africa today. But the onset of ice in the Northern Hemisphere changed all of that. It provoked the evolution of truly giant mammals, animals that could withstand the bitter cold and survive. It also stimulated a group of smallish primates to rapidly change into humans, making us among the latest of species to have evolved.[8]

Is the past a preface? The most profound ecological disasters during the history of Earth were affected by three different causes. Two of them will never occur again. The first was when oxygen appeared in the oceans and air for the first time and, subsequent to that, when oxygen levels either surged or dropped precipitously. The second was the times of such cold on Earth that huge areas of the continents were covered with ice up to a mile thick. Yet, with our ever more energetic and warming sun, it is doubtful that there will ever again be such ice ages.

But the third kind of catastrophe will surely happen again. As our planet warms, the oceans will lose their oxygen, as they did multiple times over the past 500 million years. When the high latitudes warm relative to the tropical latitudes, the global ocean goes stagnant. In the past that has led to mass extinction. It can and will happen in the future.

It is useful to reexamine the vast history of life, and especially its crises, in terms of the sequence of events that Lamarck viewed as causing evolutionary change. First was the change in environment experienced by an individual organism. Second was a change of that organism's behavior. Third was a change in its phenotype, the expression of not only how its genes were used prior to the environmental change but how they are expressed postchange. The greater the environmental change, the more consequential each of these steps might have been. For example, those organisms that survived the great asteroid impact ending the Cretaceous period would have found themselves in months of night. For those animals that fed only during the daylight hours, it was either begin feeding in darkness or die. Feeding in darkness: a new behavior.

We have to consider that this three-part sequence might work on us humans as well, especially if we view behavior as having a genetic component. Many of us believe that we are already experiencing phenomenal environmental change caused by multiple effects but with perhaps the most important being the burgeoning human population and all the environmental effects that it has produced. Other changes in our environment range from the increasing level and variety of toxins in our air, water, and food, to global warming, to the violence experienced by soldiers in wars overseas, to drug wars in American streets. More diseases; a change to strict religious life (or its converse); changes in our food type; changes to the consumption of alcohol, drugs, and caffeine; changes in our attention spans produced by our new technologies; changes in our ability to travel among and thus be rapidly placed in radically different environments around the globe in less than a day—so many ways that modern society creates environmental change.

The second effect is in behavior, and many believe that we are seeing wholesale and heritable changes in the proportions of specific behavior types.

The third, the change in phenotype, might only be manifested as behavioral change. This remains a frontier of research going forward, but first we need to look at the possible role and importance of epigenetics in the history of life.

CHAPTER VI

Epigenetics and the Origin and Diversification of Life

THERE is a fundamental duality that has existed in the trajectory of Earth life since its first synthesis toward ever more kinds of life at the cellular level. From the earliest life, there has been an increase in the number of species as well as an increase in the complexity of many of those species compared to the earliest life. The same increase is seen in the number of kinds of cells in multicellular life. Yet, at a far more basic level, Earth life must have become simpler and less diverse in its fundamental aspects.

The near unity of DNA regardless of taxonomic group might be because this is the only kind of genetic mechanism that works. More likely, however, is that there were originally many kinds of chemical assemblages that were capable of the three processes that NASA uses to define "living": being able to harvest energy from an external environment, being able to reproduce, and being able to evolve in order to survive environmental changes as well as colonize environments different from those where the first life came about.[1] A rough analogy can be made with the varied kinds of shapes and propulsion of early airplanes. Over time, however, only a few of the amazing early diversity are left. The unification came through biological mechanisms common to single-celled life that caused entire chunks of DNA to be inserted either by other single cells or, perhaps more frequently, by viruses. In minutes or less, hundreds or maybe thousands of genes that were not previously present became incorporated into the invaded cell. The new organism is not only a new *species*; it might be the start of an entirely new *family* as well. This is not Darwinian evolution. It is radical, large-scale change in a life-form. It is Lamarckian.

As discussed earlier, the frequency of the invasions and hijackings of microbial species by the insertion of long ropes of DNA is called lateral gene transfer (LGT); the significance of this is that what we call the "tree of life" is itself but a chimera because of the omnipresence of LGT, as first pointed out by W. Ford Doolittle.[2] It was (and remains!) so common that bacteria and their ilk devised a biological means of tracking down and neutralizing the newly inserted DNA, the CRISPR-Cas9 method already described.[3] In movie terms, this methodology is like identifying and defending against an Invasion of the Genome Snatchers. The tiny, already rod-shaped bacteria (or spherical, or corkscrew shaped, as all bacteria take one of these three body shapes) became something else. *Morphologically*, they still looked like their pre-invasion selves, but internally (and especially genetically) these microbes had new genetic instructions coming from many new genes that were inserted into them. They were now hijacked into a novel genetic future. Some very clever humans in this century realized how prevalent LGT was, and remains, and also discovered that bacteria evolved their own kind of defense—a "hunter-killer" means of finding and destroying the newly inserted genes before they could hijack the entire bacterium and its life functions and genetic future.

DEFINING LIFE, AND FUNDING RESEARCH INTO ITS ORIGIN

There are a lot of tricky cases concerning whether something is "alive" or not,[4] and this once again is not only related to the question of "What is life?" but to defining what being alive really consists of. These hard-to-understand cases concern single-celled parasites (such as *Giardia*[5]) and the complex and organized assemblage of proteins and nucleic acids we call viruses. Is a virus alive when it is outside a living cell? Is the intestinal parasite *Giardia* alive outside of the digestive system of its multicellular host? Are the single-celled bacteria thrown upward into near-outer-space altitudes in the high Earth atmosphere alive in the cold near vacuum of that atmosphere? These are hard questions. Life and nonlife; living and dead: We usually look at these two attributes as opposites. But increasingly it appears that they are but the extremes of a chemical and energy continuum for what we call the "simpler" forms of life, such as the single-celled microbes, and even creatures as complex as the tiny animals known as tardigrades, or "water bears," recently put into the category of the Earth life that is hardest to kill.[6] Tardigrades can be frozen to death and then thawed back to life. Life to nonlife and all in between, so unknown that even language fails us in giving us words for the "in between." And living or dead is not separated simply by a difference in chemistry and composition. The alive-to-dead continuum has the fourth dimension involved as well: time. The dead sometimes come back to life. Like vampires—alive part of the time, quite dead the rest of time.

The NASA definition of life is quite simple: (1) Life metabolizes. (2) Life replicates. (3) Life evolves. But the more fascinating question is about life itself. Can there be life without replication? The NASA scientists, or those funded by NASA in any sort of way, say no. But perhaps that is incorrect. For early life, it may have been epigenetic mechanisms that allowed life to be bestowed in an individual early cell—life that was short-lived, and left no descendant, because there was no organic mechanism that allowed replication.

The great Carl Sagan helped narrow the nature of the last of NASA's definitions of life by specifying that life not only evolves but it does so through Darwinian evolution. This was echoed later by Paul Davies in his book *The Fifth Miracle*.[7] Davies approached the question of "What is life?" by using a different question: What does life *do*? It is *actions* that define life, according to his argument. These main actions are as follows:

Life metabolizes. All organisms process chemicals, and in so doing they bring energy into their bodies. But of what use is this energy? The processing and liberation of energy by an organism is what we call "metabolism," and it is the way that life harvests enough energy and then uses it to keep order within its walls.

Another way of thinking about this is in terms of chemical reactions. When chemistry stops functioning because energy is no longer keeping "order" within, the organism supposedly has died. Not only does life maintain this unnatural state, but it also seeks out environments where the energy necessary to stay in this state can be found and harvested. Some environments on Earth are more amenable to life's chemistry than others (such as a warm, sunlit ocean surface of a coral reef or a hot spring in Yellowstone Park), and in such places we find life in abundance.

Life has complexity and organization. There is no really simple life, composed of but a handful (or even a few million) atoms. All life is composed of a great number of atoms arranged in intricate ways. It is the organization of this complexity that is a hallmark of life. Complexity is not a machine. It is a property.

Life reproduces. Davies makes the point that life must not only make a copy of itself, but must make a copy of the mechanism that allows further copying; as Davies puts it, life must include a copy of the replication apparatus too.

Life develops. Once a copy is made, life continues to change; this can be called "development." This process is quite un-machine-like. Machines do not grow, nor change in shape and or function with that growth, although new engineering breakthroughs may change that, as Google and other companies are supposedly using human-built artificial intelligence (AI) to help build second-generation AI—a machine that is machine built.

Life evolves. This is one of the most fundamental properties of life according to savants from Darwin to Davies—and, according to them, one that is integral to its existence. Davies describes this characteristic as the paradox of permanence and change. Genes must replicate, and if they cannot do so with great regularity, the organism will die. Yet, on the other hand, if the replication is perfect, there will be no variability, no way that evolution through natural selection can take place. Evolution is the key to adaptation, and without adaptation there can be no life. But if so, why must it be Darwinian and not Lamarckian?

Life is a chemical system capable of Darwinian evolution. Life according to this definition is the chemical systems that *must undergo Darwinian evolution*—meaning that if there are more individuals present in the environment than there is energy available, some will die. Those that survive do so because they carry advantageous heritable traits that they then pass on to their descendants, thus lending the offspring greater ability to survive.

THE KINDS OF EARLY LIFE AND HOW EVOLUTION EVOLVED

Early life may have been different from present life in its far higher diversity of basic forms, such as having a different genetic code or using different groups of amino acids or extracting energy in ways no longer used or perhaps even possible. Now there is one basic kind of DNA life, composed of many species—Earth life, sometimes called "Life as we know it."[8]

All present Earth life uses the same twenty amino acids. Before 3.5 billion years ago, the time when Earth life surely became unified into a single genetic code, there may have been a veritable zoo of metabolisms and genetic codes. There were surely also many life-forms that were alive as one-offs. Alive, yet without a mechanism to replicate, let alone evolve.

If this is true, we are to ask how Earth life unified into a single genetic code. Secondly, was Darwinian evolution really necessary? It seems far more likely that Darwinian evolution is the final end result of the shaping of life first by Lamarckian evolution, and the first life may have required Lamarckian mechanisms to be alive at all.

BEFORE LUCA (LAST UNIVERSAL COMMON ANCESTOR) OF "LIFE AS WE KNOW IT"

For many reasons both good and bad, 2016 will go down in human history as an astonishing year, including scientifically. "Origin of life" studies and the burgeoning experiments and papers dealing with epigenetics flourished, but as if on entirely separate scientific tracks, where most scientific research is done globally.

One of these studies led to a watershed discovery: Investigators communicated that they had discovered the almost mythical last common ancestor of all Earth life. Charles Darwin himself wrote that going back into time, sliding down the "tree of life," would lead to a basic first creature called LUCA (the last universal common ancestor), a presumably bacteria-like creature with DNA. Yet there had to be far deeper roots to that first life and the beauty of this analogy is the following: Tracing roots down shows ever more branching, ever more differentiation. That is a perfect description of life before LUCA. These recent studies suggest that LUCA, this quite primitive early rendition of Earth life, might have needed metals coming from the hydrothermal vents of early Earth to be "alive";[9] others have called it "half alive."[10] There is nothing too earth-shattering about that. Animals die quickly unless environmental oxygen is available.

There had to have been a whole zoo (but a zoo that keeps single-celled microbes) of different living organisms. Yet, if we turn around from the deepest, finest root of a huge tree, as we go up toward light and the surface of the earth, the roots coalesce. And that happened too. Some process caused a melding of the many different kinds of life: some with excellent metabolism; some that were better than others at

reproduction; some that could take in a wider variety of "food" so as to
get the energy needed for life as well as the molecules needed to build
and keep a cell running as parts of it wore out. It was the combining of
many such traits, a large number of which were or became heritable,
through the epigenetic process known as "lateral gene transfer," that
surely combined with natural selection to produce the variety of microbial life best suited to the many environments of early Earth. Thus,
by the time of LUCA—a time of our kind of life, all with the same
DNA language, using the same twenty amino acids, combined by so
many traits—there had already been a long history of life.

The discovery of LUCA added new and powerful evidence and
support for an ongoing mystery: Where did life on Earth first originate
and then live, at least our kind of life?

Two different scenarios have emerged from the many hypothesized
environments where Earth life was first constructed. Neither is where
Darwin suggested in "some warm little pond."[11] The problem with
little ponds is that they would have been highly irradiated by solar
energy, because at the time that life seems to have appeared on Earth,
somewhere between 4 and 3.6 billion years ago (depending on the
source), there was no protective atmospheric ozone layer. The two most
favored hypotheses are that (1) life first came from hydrothermal
volcanic systems deep in the sea, or that (2) it appeared in an environment where chemicals could be highly concentrated, presumably by
evaporation. Long tidal estuaries have been thought of as one possibility for this, or lagoons. The difference from now, in addition to the
oxygen-free atmosphere, was the presence of enormous tides, of larger
amplitude than in the current globe's champion, the Bay of Fundy in
Canada.

A major area of research is now trying to understand what the
minimum number of genes are that allow a cell such as LUCA to be
"alive." One set of studies[12] puts the number at 355! We would expect

the first cellular life to have few genes, but this seems like a small number indeed.

The 355 genes paint a picture of the environment that LUCA lived in, or, more accurately, the environment that allowed this living creature to survive at all. The "food" of this bacterium-like creature was hydrogen, while its other genes seem to indicate that this first Earth life had to survive in a place with high heat. It looks like LUCA lived in high-temperature hydrothermal vents deep in the primordial oceans of nascent Earth, in scalding water that would boil if it was at surface pressures today.

LATERAL GENE TRANSFER—THE SINGLE MOST CONSEQUENTIAL KIND OF HERITABLE EPIGENETICS IN EARTH HISTORY

For at least two-thirds, and perhaps three-quarters, of the time that life has existed on Earth, it was composed predominantly (or at times completely) of single-celled microbes that were prokaryotic (no nucleus or cell organelles).

In lateral gene transfer (LGT), large portions of the DNA of one archaea or bacteria is replaced through addition from a second microbe, and one of the exciting discoveries of the past decade is that LGT is not just restricted to the prokaryotes. Even humans have been evolutionarily changed by the sudden addition of new genes, injected into us by microbial vectors.[13]

This is no slow, mutation-by-mutation change. As it does now, it would have taken place probably in a short period of time—in a few hours, in fact. The LGT process is among the most Lamarckian of evolutionary processes known. Yet, the invaded microbe acquires not only a new trait but whole suites of traits. If discussed in terms of the tree of life, we are not talking about a new twig. LGT sometimes brought

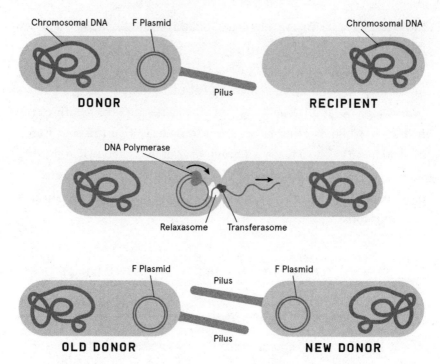

Lateral gene transfer between bacteria, where large portions of DNA in one microbe are replaced through addition from a second microbe. Barth F. Smets, PhD, Nature Publishing Group, 2005, https://media.nature.com/full/nature-assets /nrmicro/journal/v3/n9/images/nrmicro1253-f1.gif.

about so much novelty that it would have caused the new microbe, because of how new it would have been, to jump to separate families, if not become a new species. The process will also play a large role in the future of evolution on Earth, with the vector being humanity as we add new genes to plants and animals that serve as food, or in our attempts to kill weeds or insect pests in crop settings.

The principle of uniformitarianism was the bedrock (pardon the pun) of the new science of geology in the late eighteenth and early nineteenth centuries. The early "geologists" struggled to explain the presence of the many different rock types until the Scots James Hutton and Charles Lyell convincingly laid the groundwork for the principle that "the present is the key to the past": that the processes leading to igneous, sedimentary, and metamorphic rocks can be seen today.[14]

But it is not just geology that depends on the uniformitarian principle. So too does biology. It is accepted that the same processes of metabolism, reproduction, and evolution occurred in organisms all the way back to the first. Lateral gene transfer left no fossil record, as molecules do not directly fossilize (although microbes can leave a record of their past presence by leaving behind organic compounds, or "biomarkers," that could not have formed in the absence not only of life but of highly specific taxonomic branches of life).

It is not just LGT that we infer to have happened in the past. Other kinds of epigenetic processes most certainly happened in the past as well, because examples from all known higher categories of life on Earth at present demonstrate the same processes of methylation of DNA following replication, and, like animals, natural selection has probably honed many of the microbial genomes that assume methylated regions to affect phenotype, most importantly in interactions between DNA and proteins. Some of the methylated sites can produce tragic results in humans, as bacteria as common as *E. coli* can turn virulent after methylation, as can *Salmonella* and other microbes. Unfortunately, in these cases not only does epigenetic change occur, but it can be heritable.

One tragedy for the scientific reputation of Jean-Baptiste Lamarck was his use of giraffe neck elongation as a good example of his then newly thought-up process of the acquisition of traits during the lifetime of an organism. Illustrations of giraffes straining their necks upward are used in virtually every new biology textbook with the intent of making Lamarck and his theory the bull's-eye on a target of ridicule. Lamarck has become the straw-man target to be shot with Darwinian arrows. Yet, the best example of all, of course impossible in his time, would have been if Lamarck could have illustrated how a bacterium is invaded by another bacterium or a virus, leaving behind a new segment of DNA with viable genes, or the opposite: snatching genes out of the invaded organism. In both cases, when the bacterium in question reproduces, including the replication of its DNA strand with all its genes, it is functionally a different species. This is a process that would have taken minutes, not the numerous generations required by random mutation that is at the heart of Darwinian evolutionary theory.

THE DIVERSIFICATION OF THE EARLIEST
EARTH LIFE

Differentiation of the earliest life on Earth probably occurred quickly, and such differentiation into microbes with different adaptive abilities brought about the first true biological communities, where energy flows through a variety of organisms. Among the oldest of all fossils are the strange columnar objects known as stromatolites, which first appeared more than 3 billion years ago and still live in a few extreme environments today, most famously in the highly saline water body known as Shark Bay in northwestern Australia. Living stromatolites have a fascinating life cycle, as they live in jellylike slabs between thin layers of sediment, and the binding of the organic slicks of microbes with the sediment enhances fossilization into extremely hard and easily preserved evidences of life. Yet it is not just one species of microbe. The outermost layers of bacteria are forms that can photosynthesize: They contain regions in their body with chlorophyll, which by combining light and carbon dioxide can produce energy with oxygen as a by-product. Beneath them, the amount of oxygen diminishes rapidly, and the innermost layers are composed of microbes incapable of using light, and in fact poisoned by oxygen. But enough organic material and energy trickles down from the light-loving outer layer to power the entire community.

The question becomes: When did this complex community association come into being? An equally interesting and associated question is how and why did the original life on Earth, surely single cells, evolve the ability to work as an aggregate, if not as truly multicellular creatures? The individual body shapes of bacteria are composed of only three basic shapes: spheres, rods, and corkscrews. Yet today we find a huge diversity of bacteria that are giant clones, all arising from a single DNA source but becoming differentiated by alternating epigenetic regimes.

The adaptive advantage of acting like a true tissue is clear, as conditions even as commonplace as wave or current action can sweep a single tiny photosynthetic (or any other) bacterium quickly into a situation that can prove lethal (no sun, change in oxygenation of the water, and many other conditions). But when a single bacterium multiplies, eventually, such as those that can be seen in any stagnant pond as "blue-green algae" builds large clonal "colonies," it can obtain sufficient mass to better remain in place and, if it is a photosynthetic species, be able to orient toward sunlight.

If the present is at all the key to the past lives of bacteria on the early Earth, it is clear that epigenetic mechanisms were and remain vital in important life history events and traits. Among these are controlling when to replicate the DNA within a bacterium as part of the overall reproductive style of fission, where one bacterium splits into two daughter cells, each with the same DNA that has been doubled and segregated into side-by-side but separate positions correctly spaced to allow successful fission. This is no mean trick, as—unlike in eukaryotic strands, familiar as straight chromosomes—bacterial DNA is packed as one long (double helix) strand that forms a loop.[15] There are no ends. As if this were not complicated enough, bacteria have second groupings of DNA in small rings known as plasmids. Both have to be replicated for the bacterium to reproduce as a second copy.

All organisms need to constantly repair DNA. This is over and above the other "chores" of just staying alive. Such complicated and long molecules degenerate from various environmental conditions. Foremost in degrading DNA might be radiation, or heat, or other conditions outside of the adaptive limits of the organism.

Bacteria use both heritable and non-heritable DNA states.[16] In fact, they show three states of DNA: "non-methylated," or DNA that has undergone no addition of methyl molecules; "hemimethylated," where the DNA has some sites with methyl but not the full complement; and (fully) "methylated." The hemimethylated DNA is not heritable, but it has to occur for DNA repair to occur.

Among the heritable forms of DNA methylation in bacteria is the ability to alternate gene expression states, sometimes known as phases. In this process, successive generations have a binary on-off of specific gene expression. This generational change between a gene actually doing something and doing nothing (being turned off) can then be affected by natural selection. But evolution then does take place, and it was first instigated by epigenetic mechanisms that were then integrated into natural selection. Complicated? Yes. But who believes that Life is ever simple?[17]

Ultimately, the presence of patterns in the way bacterial DNA is methylated appears to have adaptive uses. The way that many microbes use methylation is a rough parallel of how we animals use "gene imprinting" adaptively. When the methylation patterns are heritable, they become almost an adaptive memory of the environmental conditions and especially the metabolic conditions within the parent microbe. Because bacteria reproduce so quickly, it is a good, safe adaptive bet that how the parent optimized its metabolism in the face of then current environment conditions will also be appropriate for the newly reproduced.[18]

THE EPIGENETIC ASPECTS OF THE MARGULIS ENDOSYMBIOSIS THEORY

The microbial world was the product of the first major diversification of DNA life. There were very few morphologies in that world, as all microbes (bacteria and archaea), as noted earlier, have but one of three basic shapes: spheres, rods, or spirals. While many kinds of microbes produce vast colony-like organisms, even the most complex of these, such as stromatolites, are still composed of only these simple body shapes. To produce life that is more complicated, life as complex as higher plants and animals, two great leaps had to be made. The first was the "eukaryotic" cell, a level of complexity that necessitated a larger (but still single) cell that had within it smaller cell-like components known as organelles. They were indeed "cell-like," because they came from cells—cells that were engulfed and then enslaved.

The endosymbiosis theory[19] was the product of one of the great late twentieth- to early twenty-first-century biologists, Lynn Margulis. It was Margulis who first detailed a model of how certain microbes engulfed others of entirely different species during their lifetimes. Eventually—and here is where the great complexity comes in—these captured cells became part of a larger cell itself that reproduced the now genetically integrated organelles, such as the nucleus, mitochondria, and plant chloroplasts, among others. The process was Lamarckian, which Margulis herself accepted: "According to present-day neo-Darwinian evolutionary theory, the only source of novelty is claimed to be by incorporation of random mutations, by recombination, gene duplication, and other DNA rearrangements. As is emphasized by those using the term symbiogenesis, symbiosis analysis contradicts these assertions by revealing 'Lamarckian' cases of the inheritance of acquired genomes."[20]

Margulis's ideas centered on the "how" of evolutionary novelty, and for the microbial world of more than 3 billion years ago, the evolution of the eukaryotic cells was a path to enormous novelty. The process was soon accepted by the major thinkers of the time, such as John Maynard Smith, who noted: "The relevance of symbiosis is that it affords a mechanism whereby genetic material from very distantly related organisms can be brought together in a single descendant."[21]

Endosymbiosis as a Lamarckian event was of Margulis's many prescient understandings. She viewed it as a far more powerful means of formation of evolutionary novelty than Darwinian evolution. It has been noted that natural selection acting on a newly mutated gene cannot from this create a new gene. In endosymbiosis, however, Margulis posited that the process produced the merging of thousands of individual genes, each of which had already run the gauntlet of mutation to usefulness and applicability within the genome of some species. In this way the potential for change would have been orders of magnitude more than the gradual, single mutation-by-mutation change of Darwinism.

How fast might such an engulfing of one species by another have happened? While we have no time machine, we can study modern organisms because this phenomenon continues today. The process is called *phagocytosis* and involves the "eating" of one cell by another. Actual predation. A process that could have happened during any given hour on any given day that in turn would cause great evolutionary changes (in some cases) in subsequent descendants of the voracious cell eater in question. Some have called this theory (and it does stand as a theory now, because there has been so much research that has been unable to falsify it) the "fateful encounter." As noted by my esteemed colleague Nick Lane in his book *Life Ascending*, "The fateful encounter theories are all essentially non-Darwinian, in that they don't posit small changes as the mode of evolution, but the relatively dramatic origin of a new entity altogether . . . The implication is that there was something about the union itself which transformed the arch-conservative, never-changing prokaryote into its antithesis, the ultimate speed junky, the ever-changing eukaryote."[22]

And was it the *purpose* of the larger cell to engulf the smaller in these fateful encounters? Probably, but also coupled with situations where it was the "purpose" of the smaller cell to invade the larger cell: the invasion of a larger cell by a smaller for purposes of the smaller *at first*.

Once inside, the bacterium within a bacterium became locked in place for the benefit of both. So how did they fuse into a single creature with unified DNA? The explanation seems to be that some genes (stretches of DNA) are called "jumping genes." They copy themselves in snippets of RNA and then jump back onto the much larger DNA. But in some cases they jump not back onto their original DNA but onto the DNA of the microbe that engulfed them. We see this process as the means by which mitochondria, the small intracellular power stations that run cell metabolism in us eukaryotes, have over time had almost their entire original DNA genes "jump" into the DNA held in the nucleus of their parent cell.

Here is a wonderful case example from the University of California, Berkeley, website called Understanding Evolution:

> In 1966, microbiologist Kwang Jeon was studying single-celled organisms called amoebae when his amoebae communities were struck by an unexpected plague: a bacterial infection. Literally thousands of the tiny invaders—named x-bacteria by Jeon— squeezed inside each amoeba cell, causing the cell to become dangerously sick. Only a few amoebae survived the epidemic. However, several months later, the few surviving amoebae and their descendants seemed to be unexpectedly healthy. Had the amoebae finally managed to fight off the x-bacterial infection? Jeon and his colleagues were surprised to find that the answer was no—the x-bacteria were still thriving inside their amoebae hosts, but they no longer made the amoebae sick. There were more surprises when Jeon used antibiotics to kill the bacteria inside an amoeba—the host amoeba also died! The amoebae could no longer live without their former attackers. Jeon discovered that this was because the bacteria make a protein that the amoebae need to survive. The nature of the relationship between the two species had changed entirely: from attack and defense to cooperation.[23]

What is not seen here is the next step: that the amoeba in question now reproduced and built the small bacteria within by co-opting the genetic code of the bacterium in question. These observations are definitely epigenetic: A change in environment (invasion by microbes) took place. There was a change of "behavior"—in this case, the presence of the bacteria became necessary for the larger cell to live. The last step would be codifying this genetically.

By being able to engulf the far smaller prokaryotic cells at will, the early (and far, far larger) eukaryotic cells did not have to wait for the long, slow mutation-by-mutation kind of evolution, with each single mutation then acted on by the crucible of natural selection. Instead, they could "eat" an entirely new library of genes and assimilate them by forms of lateral gene transfer. In mechanism this is different from the lateral gene transfer described earlier. The engulfing, and then jumping, genes gave the eukaryotes a rapid means to try out all kinds of new genomes, which meant new kinds of life. Natural selection would act on these new life-forms, but it was Lamarckian epigenetic processes that led to the rapid change in body plans and physiology.

Prokaryotic cells are conservative but amazing chemical factories. When encountering a sudden environmental change in the chemistry of the water they are in, they try to change the water. Eukaryotic cells do not try to change the water. They change themselves to be able to live in that water by building new body parts, and they do this by eating the tiny bacteria around them.

Complex multicellular life was a relatively late development on Earth. That the world has been dominated by prokaryotic life, and that this is the seed that generated the Tree of Life, makes the case that much of the evolutionary history of Earth life has been dictated by Lamarckian processes, aka heritable epigenetics.

CHAPTER VII

Epigenetics and the Cambrian Explosion

O N E of the astonishing aspects of the epigenetic revolution is that it has been virtually ignored in summaries about the history of life on Earth. For decades, the religious extremists known as creationists have criticized evolutionary theory on two things: complexity (too much to have come from random genetic mutations) and rates (from the fossil record, species appear "too quickly"). Of special concern to them has been the Cambrian explosion, the interval of time and the biological results of a half billion years ago, when the major body plans of animals now on Earth appeared rapidly in the fossil record.[1] Yet the processes of epigenetics can explain away the creationist complaints about the history of life.

Stephen Meyer, a member of the intelligent design contingent, has repeatedly noted the "impossibility" that so many animal body plans appeared in the Cambrian period through Darwinian explanations alone. In this I agree with him. It's just not fast enough, he argues.[2] He is right. But the Deity he worships should be Lamarck, not God.

The creationist's criticism is that Darwinian mechanisms, most notably natural selection combined with slow, gene-by-gene mutations, can in no way produce at the apparent speed at which the Cambrian explosion was able to produce all the basic animal body plans in tens of millions of years or less. Yet the evidence of even faster evolutionary change is all around us. For example, the way that weeds when invading a new environment can quickly change their shapes. The question is not whether there was epigenetic formation of new body plans (that were then epigenetically sent forward in time through heritable epigenetics) but whether traditional Darwinian evolution had much of *anything* to

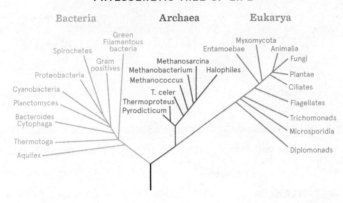

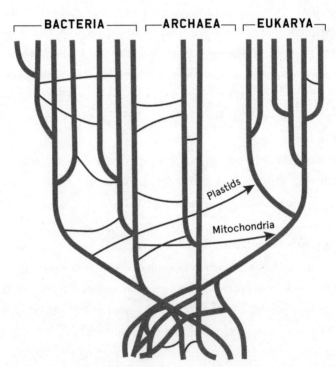

Top: The "Standard" Phylogenetic Tree of Life depicts evolutionary changes vertically, as traits changing through time. Bottom: The Lamarckian Phylogenetic Tree of Life depicts evolutionary changes both vertically and horizontally, as traits changing through both time and epigenetic transfer. Peter Ward, *Life As We Do Not Know It* (New York: Penguin Viking, 2005).

do with the multiple times where rapid evolution clearly took place. One thing is certain: there is no need to invoke the supernatural.

It is posited here that four different epigenetic mechanisms presumably contributed to the great increase in both the kinds of species and the kinds of morphologies that distinguished them that together produced the Cambrian explosion as we currently know it: the first, the now familiar methylation; second, small RNA silencing; third, changes in the histones, the scaffolding that dictates the overall shape of a DNA molecule; and, finally, lateral gene transfer, which has recently been shown to work in animals, not just microbes. It does not take place gradually, and it does not take millions of years, or even millions of seconds. Unfortunately, these are not processes that leave fossils in hard parts. But observation of all of these processes in extant organisms (especially through modern genetic studies) provides reasons that they can be assumed to have been at work in deep time. As noted in the previous chapter, of the four, it is lateral gene transfer that seems to be the most Lamarckian of evolutionary changes, as the process as observed today is quickly done and causes lasting change in the genome during an organism's life.

LOCOMOTION AND THE CAMBRIAN EXPLOSION

So much has been written about the Cambrian explosion that little new seemingly can be added. But it is surprising to find little to no inferences of how epigenetic processes may have been involved in the two major changes that seem most important as a result of this time interval during the Cambrian period, from about 544 million to about 500 million years ago. The first was the great rate at which entirely new body forms appeared. Not just single changes, but entirely new anatomies of animals that were then genetically stabilized to form the taxonomic categories called phyla. As is well known, all the animal phyla appeared over this interval. The great rapidity was the scourge of Darwin. He was presented with the fossil record, indicating that trilobites, highly complex animals, were the *first* animals.[3] Darwin then had to somehow establish the series of trilobite precursors. Where were all those intermediate forms in the fossil record?

The fossil record is far better known now. We know that trilobites actually appeared in the latter stages of the Cambrian explosion and were preceded by many arthropods.[4] Yet the fundamental problem still lingers. How could so much evolutionary change have occurred so quickly?

Charles Marshall of the University of California, Berkeley, long ago noted that one of the salient changes evolved in early animals was the ability of actual locomotion.[5] Movement by legs, fins, undulations, a cornucopia of morphological and concomitant physiological changes from inside to outside. The preceding and perhaps first animals, the strange Ediacarans, were motionless, sitting their whole lives on the same patch of seabottom. But the base of the Cambrian system is marked by the first "trace fossils," which are nothing less than the evidence of movement. Movement is behavior, itself a brain function. Recently, one of the great doyens of the field of epigenetics, Eva Jablonka, co-authored with Simona Ginsburg a heretical yet highly logical and probably correct suggestion[6] that it was a breakthrough in learning, allowed and passed on through epigenetic processes, that was a major aspect of the reasons producing the rapid evolutionary changes seen in the Cambrian explosion.

In the necessarily dry scientific prose required in published science articles, Jablonka and Ginsburg put forward this radical new idea, revolutionary in evolutionary importance. Heretofore it has been mostly paleontologists who have tried to understand the history of animals from the clues given in the rocky and (to Darwin) maddeningly incomplete fossil record. But this was a new take on the Cambrian explosion coming from scientists versed in the possibilities of explanation offered by an understanding of epigenetics. For the first time, it was posited that a new marriage existed between stress hormones that allowed newly evolving animals to exploit increasingly complex behavior that was aided by new and exquisite sense organs—including, for the first time, efficient eyes. Jablonka and Ginsburg called this "learning-based diversification."

This says that changes of behavior by both animal predators and animal prey began an "arms race" in not just morphology but behavior. Learning how to hunt or flee; detecting food and mates and habitats at a distance from chemical senses of smell or vision, or from deciphering vibrations coming through water. Yet none of that would matter if the new behaviors and abilities were not passed on. As more animal body plans and the species they were composed of appeared, ecological communities changed radically and quickly. The epigenetic systems in animals were, according to the authors, "destabilized," and in reordering them it allowed new kinds of morphology, physiology, and again behavior, and amid this was the ever-greater use of powerful hormone systems. Seeing an approaching predator was not enough. The recognition of imminent danger would only save an animal's life if its whole body was alerted and put on a "war footing" by the flooding of the creature with stress hormones. Powerful enactors of action. Over time, these systems were made heritable and, according to the authors, the novel evolution of fight or flight chemicals would have greatly enhanced survivability and success of early animals "enabled animals to exploit new niches, promoted new types of relations and arms races, and led to adaptive responses that became fixed through genetics."

That, and vision. Brains, behavior, sense organs, and hormones tied the nervous system to the digestive system. No single adaption led to animal success. It was the integration of these disparate systems into a whole that fostered survivability, and fostered the rapid evolution of new kinds of animals during the evolutionary fecund Cambrian Explosion.

UPGRADING SENSES IN CAMBRIAN ANIMALS

One of the most critical aspects of animal success concerns brain function and intellect. One of the many processes of our brain, and those of our many 500-million-year-old vertebrate ancestors, is learning. But to accelerate learning, there needed to be upgrades in sensory input.

One of the interesting hypotheses about the Cambrian explosion was that the predators, and eventually their prey, evolutionarily discovered the immense advantage of vision.[7] While in murky waters today, many fish rely on their lateral line systems to detect the presence of potential predators from vibrations in the water, in even murkier river water, highly specialized electric eels and their kin use electric charges for the same effect. Even so, vision remains a highly useful sensory tool to find prey and mates, and to avoid becoming prey. But vision alone is useless unless there is an advance in neural capacity and brain power to unravel what is seen, let alone "see" what is being seen.

But even this is less than optimal without the ability to intellectually *profit* from prior experience. To learn, in other words. Jablonka and Ginsburg have thus made the highly interesting hypothesis that increased learning ability in early vertebrates *itself* became accelerated as well as channelized evolutionary change. First learning, and then in a positive feedback loop, better vision, better interpretation of what is seen, and, most important, change in behavior. As Lamarck so long ago hypothesized: Behavior changes first, which later leads to morphological changes.

This kind of feedback loop certainly did not end with the Cambrian explosion; that was just the beginning of rapid and radical changes in brain ability. When vertebrates finally crawled onto land to stay, some 350 million years ago, the same kind of process probably helped drive the new kinds of eyes needed for vision in air, the new kind of hearing, and, with those two, increases in communication. But these had to be tied to internal signaling systems using chemicals (hormones) that themselves produced fear or flight, coupled to cooperative behavior, and later still altruism, leading to the evolution of intense emotions and emotional bonding, all accelerated by symbolism.

A thought, a vision of a mate in an increasingly symbolic fashion, coupled to the already advanced hormone systems, saved lives. Birds luring predators from mates and young. Animals bringing food back for young. When symbolic intellect was coupled to intense bursts of serotonin, whereas danger to those same family members caused cortisol and adrenaline to spike, and packed in symbolism of family and tribe and loyalty, natural selection kicked in. Survival of a population sometimes depended on the sacrifice of some of its members. Such altruism had its origins long ago, perhaps not yet in the Cambrian. But that was the time when we vertebrates made our first appearance in the fossil record with the small, worm-sized and worm-shaped lancelet-like creatures that Stephen Jay Gould gloriously wrote about in *Wonderful Life* (itself rich in symbolism from cultural to deeper emotional context about chance in the history of life, and chance in the history of our own lives, Steve Gould once told me). That animal, by the name of *Pikaia*, would not have been present at all as the deep ancestor of humanity without epigenetic processes.

Eva Jablonka and Simona Ginsburg were not the only pioneers in thinking about the Cambrian explosion in the context of epigenetic change. Another ambitious paper[8] also proposed that a basic epigenetic mechanism was the "trigger" that precipitated the Cambrian explosion, but from a quite different anatomical point of view than that of Jablonka and Ginsburg.

The author of this work, Chris Phoenix, pointed to a major biological difference between the first two animal phyla to appear on earth, sponges and cnidarians, the latter comprising corals, anemones, and jellyfish (although new work[9] suggests that jellyfish-like ctenophores may have predated the cnidarians). While these early groups are indeed "multicellular" animals, neither sponges nor jellyfish have cells that differentiate in a one-way (unidirectional) manner in the way that all subsequent animals do.[10]

In the more "advanced" animals that came after sponges and cnidarians, once a nerve cell becomes a nerve cell, or a muscle cell becomes a muscle cell, it stays that way through life. The major changes that allow rapid, directional locomotion—something a mobile predator would need—requires a bilateral symmetry, and a bilaterally symmetrical body plan gives the animal a head, a tail section, and directionality. From there the head region can become stocked with sensory organs, which requires some kind of brain, all of which needs cellular differentiation. This kind of extreme specialization of cells *demands* some kind of epigenetic control. Most animals have what is known as an organ-level of complexity. Organs are composed of specialized tissues which are in turn, composed of various specialized tissues. Yet how does any organism go from fertilized egg to the complexity of animal? Cells in organs have the complete genetic complement, yet most of those genes are not needed. For example, a liver cell does not need the genes required to build red blood cells. Turning large blocks of genome on or off can be done slowly, one mutation at a time, or as huge chunks of change, through methylation. In one short article, Phoenix provided a new perspective.

These two scientific papers came in 2009 and 2010. From there the floodgates relating to epigenetics and the Cambrian explosion opened, yet none of this has made it into the textbooks thus far.

HOW TO BE A FISH IN A SLOW- (OR NO-) MOTION WORLD

Of all the creatures that appeared in the Cambrian explosion, none are of more interest to us than our own group, the chordates. While several of the earliest evolved chordates had some strange shapes, quite early in the history of the group it was one kind of anatomy that caused runaway diversification: a sleek, fusiform body built around a linear "backbone," producing the highly cephalized (prominent head in front) shape familiar to us all: the fish shape.

In the world of the Cambrian period, amid the Cambrian explosion, there were far more undersea animals that were sessile during most of their lives than there were those that were motile. While many of the eventually sessile animals—such as sponges, anemones, many mollusks, all brachiopods and bryozoans, tunicates, and tube worms—did spend some time immediately after hatching as passively drifting, microscopic larvae among the marine plankton, for most of their lives they never moved. And because they never moved, few had good sensory organs or even heads. One of the great advances in animal design was the evolution of the bilaterians: organisms with a head, a body, and a tail that is bilaterally symmetrical from side to side, but not front to back.

The advent of this front-and-back axis, of being bilaterally symmetrical in a body where a head was encrusted with newly evolved sensory apparatuses, allowed a new kind of living: predation. But until the very end of the Cambrian, only the most advanced of arthropods had such body plans and thus posed any competition to a newly evolved ability that they shared with the earliest chordates: the ability to rapidly swim, either to chase prey or escape from prey. Our deepest ancestors inhabited bodies yet without bone, and their closest living relative was a creature known as the lancelet, or amphioxus, as it is known among the scientific community. But the arthropods had something that these small, simple chordate animals did not: jaws, as well as appendages such as claws that acted as offensive weaponry in the war of survival. Soon thereafter (in the late Cambrian period), the arthropods were joined by another predatory group, probably also carnivores on primitive fish: the cephalopods. It behooved the early fish to have the best body possible for escape. The fishlike shape still present rapidly appeared, and it was allowed and abetted first by a horizontal nerve cord running from head to tail, which was later surrounded in the bony fish by a backbone.

The ability of the early fish to move through the sea in a horizontal line was nearly unprecedented, and they were more streamlined than the bulkier arthropods. No matter that the still-small fish were unable to bite or ingest anything beyond microscopic prey; they were like present-day tadpoles, living off bacterial and algal scum. These first of our group needed a bevy of new anatomical adaptations to survive, such as the ability to see the predators coming for them and the ability to swim rapidly to escape, for perhaps most driving of natural, selective factors was the need for an ability to avoid becoming meals of the diverse and hugely successful arthropods and cephalopods.

While there were predators among the Cambrian arthropods were mud-grubbing, slow-moving detritus feeders and no threat to other animals: These were the trilobites. They subsisted on organic-matter-rich mud straining out the merger nutrients to be found there. The ancestors of scorpions and spiders lived underwater, as did arthropodan groups that at the time were the largest animals in the world, and surely the most fierce and predatorily successful: giants such as the almost lobsterlike *Anomalocaris*, a nearly six-foot-long terror with two eyes (each with 8,000 lenses) and a long, undulating swimming body that routinely attained the fastest swimming speeds the world had known to that date, and indeed an animal that if unleashed in the modern world would probably more than hold its own.

Three kinds of rapidly moving predators emerged from the Cambrian—the arthropods, the cephalopods, and our ancestors, the chordates. All were swimmers. But of them, it was the streamlined, vertebrate body plan that still exists in the myriad fish of the present day, a body plan with its brain-bearing head and long, slim body perfect for swimming. Yet this aquatic design was also composed of parts that lent themselves to be evolved into land life as well. While the arthropods also made it onto land as well, their design could never reach sizes threatening large land vertebrates for on land the arthropod exoskeleton design is quickly crushed by gravity when even approaching the size of a large dog. As for the cephalopods, they never got out of the sea.

THE HORMONES OF THE CHORDATES—UNSUNG
KEYS TO OUR SUCCESS?

New research[11] has shown that the most advanced hormonal stress system ever evolved initially appeared in the first fish, perhaps as early as 500 million years ago. Yet, old as it is, these same hormones can be found not only in the still-living descendants of primitive Cambrian fish, the loathsome lampreys and slime eels, but in all of us chordates, including humans. We know now that the amount of stress hormones in the blood of vertebrates, including humans, is a balance between external environmental triggers and internal physiological responses, and that our own system has been honed by 500 million years of evolution.

Yet just as important as the specific hormones evolved was the evolution of what is called the "gut-brain axis,"[12] or, more formally, the hypothalamic-pituitary-adrenal axis, shortened to HPA axis.

The HPA axis has three anatomical parts, each in a different part of the body. The hypothalamus is in the brain; the pituitary gland is just beneath it, in the brain stem; and the adrenal glands are far from either, sitting just above the kidneys. Each plays a different function.

In terms of environmental stress, the HPA axis process goes as follows, and the process is what evolutionists call "highly conserved." (The steps outlined below took place in the first vertebrates and take place still, although the details and complexities of the system have changed and increased.) Following a stress stimulus (or multiple stimuli), neural cells in the hypothalamus synthesize small molecules called neuropeptides that have a very specific function: They are the trigger for the hormone-producing pituitary gland to produce a second, larger molecule (called adrenocorticotropic hormone, or ACTH) that is released into the blood or other circulation system of the body. When these ACTH molecules finally arrive at the adrenal glands, they bind to receptor sites, like keys that fit into a keyhole, and unlock (as well as trigger) the formation of powerful corticosteroids.

Steroids have a bad rap in a human context. Even small quantities can radically affect human performance, which is why they are banned from use among baseball players, Olympic athletes, and Tour de France bike racers, among others.

Corticosteroids come in two different functional groups in mammals, and they affect mineral balances in the body (such as salt content and hence blood pressure, among many other purely physiological responses) as well as providing fast-acting chemical wake-up calls to the entire body in some environmental situations (such as being chased and eaten, among many other possibilities). They also play a major role in the fetal development of chordates.[13]

Corticosteroids are crucial following fertilization of a viable egg, and especially in influencing the proper growth of the heart, lungs, and brain. But another role of this powerful group of hormones is in allowing the function that sets off the chordates from animals in most other phyla: They control the levels of minerals such as calcium and phosphates needed for producing movement, be it rapid or long term. Phosphates are necessary for the energy-producing mitochondria to replenish stocks of adenosine triphosphate (ATP), the small chemical batteries that power muscle functions as well as other energy-needing functions of any body. For example: An early jawed fish was attacked by a large predatory sea scorpion. There was a chase, perhaps a significant chase, in which the fish was flat-out swimming for its life. It eventually escaped, and it was made to feel immediately hungry. The fish ate to restock levels of blood sugar as well as elements such as calcium, phosphorus, potassium, and even iron (for hemoglobin in blood) that were paramount after exertion. Some steroids stimulate the "hunger" sense and thus provoke a feeding response. Feeding response is a behavior. But so too was the escape response in the first place, also stimulated by hormones flooding into the body.

The corticosteroids also have another crucial aspect: They are self-regulating by what is known as a negative feedback system. Too much of any good thing is a bad idea. No vertebrate can function long with the HPA system switched perpetually to "on." As more and more of the corticosteroids flood the body, they begin to be taken up by receptors, which neutralize their functions, their many, many functions. The higher the hormone level, the faster (and/or more numerous) these absorbing sites begin to function to clear the animal of the high levels of corticosteroids.

As in so much in life, the first step in constructing the HPA system during the growth of an animal is regulated by specific genes. Natural selection has long acted on these genes, and the fact that the HPA system has been around for so long in our own lineage (a half billion years) indicates how useful it is in so many ways, from building bodies to keeping mineral levels at the correct concentrations to provoking responses to environmental change.

This is where the intersection with epigenetics can be identified. Environmental change during the life of an organism, be it a microbe or a whale, can stimulate change that in some cases becomes heritable. Science is just beginning to recognize how often, and in which species, and especially in which situations these changes actually take place.[14]

The Cambrian explosion set the stage for the domination of land and sea (in terms of the largest animals in either) by being the interval and event that produced all of the animal body plans, or "phyla." None have been produced since, not even after mass extinctions subsequent to the Cambrian explosion. For the most important of the phyla with regard to domination of land, the key to our success has been epigenetic processes working toward ever greater intelligence, behavior, and the intersection of hormone systems tying the gut and brain together. But of the many processes that combined, only recently has the role of stressors and epigenetics come into recognition as being involved at all.[15]

CHAPTER VIII

Epigenetic Processes Before and After Mass Extinctions

M ASS murder simultaneously fascinates and horrifies, and that simple reality perhaps explains our fascination with the mass extinctions of deep time, the multiple occasions over the time since animals emerged when the majority of species capable of fossilization "suddenly" disappeared from the fossil record. Interest increased with the sensational discovery by geologists from Berkeley (Walter and Luis Alvarez, Frank Asaro, and Helen Michel) that the most famous of all prehistoric creatures, the beloved dinosaurs, apparently disappeared in haste simultaneously with the impact of a large (ten kilometers in diameter) asteroid with Earth, an event dated to around 65 million years ago. As every schoolchild knows, mammals then crawled out of our foxholes and out of our rat-sized body plans to quickly evolve into planet-spanning and (at least on land) ecosystem-dominating larger animals.

The paradigm of large body impact as a cause of mass extinction, formulated in 1980, was by the close of the twentieth century the favored hypothesis not only for the "K-T" mass extinction (unfortunately changed subsequently, surely to the sorrow of PR people, to "K-Pg," for Cretaceous-Paleogene) but for the other four of the Big Five mass extinctions of animals. By 2010, however, it was clear that instead of being but one of many, the K-Pg impact 65 million years ago was unique in creating a global catastrophe eliminating more than 50 percent of species capable of fossilization.[1] Even more extensive than the K-Pg event was the Permian-Triassic mass extinction 251 million years ago.[2] In the end, which extinction killed more?

Paleontologists have tried to sum up the exact number of taxa killed off in the mass extinctions, except such exercises are ultimately self-defeating, because my fraternity cannot agree on how to count the dead. Should it be the total number of taxa killed off? (And even here there is disagreement: Should it be the total number of families, genera, or species?) Or should it be the percent of individual organisms that died off as the primary measure of relative catastrophe, much as us humans rate the lethality of our wars by the body count? There is no consensus. It is like asking which was more catastrophic: World War I, in which fewer humans died globally yet in the context of the total global population a higher percentage of the human population was killed off, or World War II, during which the human population was larger?

In our 2015 book, *A New History of Life*, Joe Kirschvink and I identified ten such events going back to billions of years ago. The so-called Sixth Extinction of today is way more than the sixth. We identify it as the tenth.[3]

One of the most perplexing ongoing questions faced by the field of paleobiology concerns the number of species present in any time interval, and it also concerns the number of species killed off during the various mass extinction events. About the number as well as the *percentage* of species that die out in any million-year period during times of "normality," when there were no major catastrophes. Species do go extinct, most commonly during the normal (no-catastrophe) times at a given rate, and the cause is usually either competition from better-adapted species or from predation from some new and highly effective predator.

But the problem with such studies deals more with the math used than with the actual events. Were more soldiers killed in World War I or II? A smaller percentage of the armies of World War II were killed off than those of World War I, but the armies were larger. Was a higher *percentage* of the total size of the armies killed?

DARWINIAN VS. LAMARCKIAN EVOLUTION

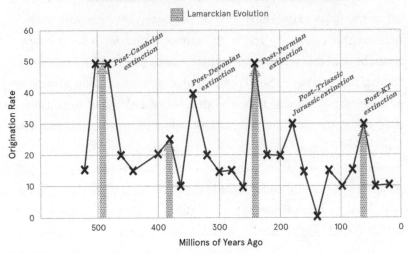

Lamarckian (epigenetic) evolution helps to explain how large numbers of new species appeared following mass extinctions. Based on Peter Ward and Ross Mitchell, "Epigenetic vs. Darwinian Time," abstract with program, Geological Society of America, Penrose Conference City of Apiro, Central Italy, September 25–29, 2017.

The answer remains unsatisfactory, since our numbering the dead depends not only on the level that we count but also on the absolute number of species (or genera, or whatever taxonomic level is serving as the global population) that were present. The first biologists who attempted to assess global diversity through time suggested that since the first appearance of animals, the number of them has increased over time (except for the short-term reduction from the mass extinctions). If so, then a 90 percent extinction some 250 million years ago might, in fact, have killed off far fewer species than a 70 percent extinction some 65 million years ago, assuming the number of species doubled between those two times.

A different and perhaps more clarifying way to look at these global events over the past 600 million years is to concentrate not on the number of species but on the number of body plans that died out. Pheno-type (shape and anatomy) is driven by genotype plus epigenome. How many unique body plans of animals and plants died out and were replaced? Or not replaced but superimposed ecologically by new species that not only had entirely new kinds of body plans but had new kinds of "jobs" in ecosystems? Prior to the Early Cretaceous (about 130 million years ago), there was no need for pollinating bees because there were no flowers at all. The job of existing on flower pollen did not exist, just as the job of software engineer did not exist prior to the 1970s. The expansion of life often produced new kinds of animals in response to entirely new kinds of body plans. Why evolve a butterfly before flowers have been invented?

THE PARADIGMS OF MASS EXTINCTIONS AND THE ROLE OF EPIGENETICS

So many books have been written about mass extinctions that new boilerplate descriptions seem superfluous. At the present time there appear to some of us studying mass extinctions that there have been only three main kinds of mass extinction causes, at least during the time of animals: (1) The rarest were impacts of giant asteroids or comets from space. (2) Greenhouse mass extinction events, where large-scale volcanism caused rapid global warming that in turn caused oceanic starvation of oxygen because of the reduction of pole-to-equator temperature. In this model of mass extinction, warmer tempera-tures reduce or stop ocean currents, depleting oxygenation via the thermohaline conveyer belt currents. (3) Extinctions caused by cold and glaciation.[4]

New research should be focused on the role, if any, of epigenetic processes in these three extinction-causation hypotheses. Clearly, epigenetics did not produce or control any massive, interplanetary snowballs that hit Earth, nor launch asteroids at us, nor cause flood basalts. But what about any role of epigenetic processes in the wake of such events? Rocks from space killed off huge swaths of Earth's biota and triggered wholesale changes in communities of organisms by rapid evolutionary change.

KNOWN EPIGENETIC EFFECTS PRODUCING EXTINCTION

The most consequential mass extinction of the past 500 million years has attracted many nonscientific names, including the Great Dying and even the Mother of All Mass Extinctions.[5] Estimates of its death toll depend on which taxonomic level one looks at. A famous estimate by the late David Raup, a University of Chicago paleontologist, was that over 90 percent of genera went extinct—and every genus is normally made up of multiple species.[6] Yet Raup's prediction came from the 1970s, and few professionals have attempted a rigorous examination since. It is clear that most species capable of leaving a fossil record died out. But perhaps there are many times more bacterial and viral species than currently accepted.

Global diversity today is listed as close to 2 million species, while some estimates are as high as 20 million; the majority of species would have to be microbes. Microbes, it seems, remain untouched by mass extinction at the species level. The Permian extinction may have knocked out perhaps 70 percent of animal species, but the total microbes it killed may have been far less. We animals are chauvinists, however. We tend to dismiss the microbes as unimportant.

So even if the number of species killed off is smaller than earlier believed, the Permian extinction is among the most devastating mortality events for the various kinds of life in the history of Earth. Trying to figure out its cause has been a cottage industry among Earth scientists. In the beginning of this century, another of those pesky asteroids from space was invoked, a hypothesis based on the known fact that the dinosaur-killing K-T mass extinction of 65 million years ago was largely or completely caused by asteroid impact. But this was certainly not the case for the Permian extinction, based on abundant evidence from the late 1990s up to the present day. Around 251 million years ago one of the largest flood basalt eruptions known from geologic time on Earth took place, putting all manner of greenhouse gases, including carbon dioxide and water vapor, into the atmosphere. A 2014 study[7] now adds the presence of another kind of gas, methane, which is also a greenhouse gas that can heat up the Earth.

Dan Rothman and his colleagues at MIT were able to put all of this together by combining geology, genetics, geochemistry, and evolutionary biology to arrive at a hypothesis: that epigenetic change was involved in the greatest mass extinction.

For several million years prior to the Permian extinction, the world was in one of its most biologically productive time intervals, in that the volume of plant, microbial, and animal life on Earth may have been higher than any time previous. This interval, from around 300 million to about 250 million years ago, was a time of great plant and animal success in terms of the number of species on Earth but even more so in the actual mass of life that was produced. It was a time when plants and animals thrived, and the volume of not only living material but also organic material accumulating as a result of these "boom times" for life was very consequential, and very large. Just as a warm and friendly summer produces a great quantity of fallen leaves in autumn, so too did the Permian time of booming life leave an increase in the amount of organic carbon accumulating on the planet in the form of dead plants, animals, and microbes.

The seasonal accumulation of organic material produced in this thriving world, such as fallen leaves and twigs on land as well as the dead remains of a rich Permian age plankton in the seas, carried enormous quantities of energy-rich organic compounds down to the world's ocean bottoms, from shallow to deep. On the shallower sea bottoms, the dead organic material would rot, but vast quantities found their way to lake and sea bottoms where there was little or no oxygen. Because global temperatures then were high (much higher than today's), the temperature differences between the high-latitude north and south poles and the low-latitude equatorial regions were less than those of today. Warm air masses and warm water masses move to cold. Yet, in the late Permian world, there was already little ocean circulation at the surface, and even less at depth. Today, there is vertical (with depth) circulation as well as horizontal currents such as the familiar Gulf Stream. With little wind, no Gulf Stream equivalent, and few to little motion carrying cold, oxygen-rich surface water down to the deep sea, the enormous quantities of energy-rich dead bodies from the oxygen-rich world fell to oxygen-deficient sea bottoms.

At the same time, the energy-rich bodies—leaves and other remains of the once living—became surrounded by the organic compound acetate. This stuff acted against the already low levels of oxygen-requiring microbes on the deep-sea bottoms like mothballs do to moths: Acetate stopped ocean-bottom bacteria of the time from using the fallen organic material for food. This literally would have been a boon for the other deep-sea microbes called "methanogens" (because they release methane after respiration in the near or total absence of oxygen)—if, that is, the methanogens had the genetic mechanisms to be able to use the dead organic material for food. Which they did not, at first. All the late Permian sea bottoms were accumulating masses of "reduced" or energy-rich material. It was a pile of food sitting there, but with little in the way of species able to eat it because the food was coated with the unpalatable acetate molecules.

Enter epigenetics, via the previously described lateral gene transfer. Just as in the case of the first diversification of life through lateral gene transfer, so too did the methanogens change their genetics by taking up new genes. The methanogens captured two acetate-processing genes from a totally different kind of microbe, one that was even in a different kingdom. This world-changing Lamarckian event has been dated to about 250 million years ago, and it was discovered by geneticists comparing the genomes of fifty different modern organisms. By using a "molecular clock approach," which times the antiquity of organisms by comparing similarities in their DNA, the origin of microbes capable of dealing with an acetate-polluted food source happened essentially contemporaneously with the Permian mass extinction. The newly reengineered ocean-bottom methanogens went to work. And in so doing, they produced vast volumes of methane, which is among the most potent of all greenhouse gases. The gene capture enabled the methanogens to vastly multiply in number. The methane they produced then added to the huge increase in greenhouse gases being at the same time liberated from an enormous flood basalt eruption in what is now Siberia, known as the Siberian Traps flood basalt event.

The result was a rapid heating of the biosphere, the places where Earth life lives. The new gas content both dissolved in seawater and later found in the atmosphere was deadly to all organisms that need oxygen and all organisms that die at sustained temperatures at or above about 35°C (95°F). Which is most everything that is multicellular.

The question is whether the epigenetic transfer of the acetate genes was inherited or if this happened to all of the microbes sometime in their lives and was then passed on. Microbes simply split in half by duplicating their DNA, with a complete copy to each. And they do this quickly. It seems likely that this late Permian event was a product of heritable epigenetics rather than a simple epigenetic change that died when the organism died and was not passed on with each reproduction. At the same time, what happened to the bacterium that gave up this gene? Did it die out immediately? Was this akin to a fatal parasitism? For the donor microbe, it was akin to being robbed of its genome, and not even all of it.[8]

The MIT work of Dan Rothman and his colleagues has given us one of the most concrete of all examples of epigenetics as a response to global environmental change at the mass extinction–producing level.

DOMESTICATED ANIMALS AS CLUES TO THE
AFTERMATH OF MASS EXTINCTIONS

The question of the mechanism of very rapid evolution is central to unraveling the mystery of both the Cambrian explosion and the rapid recovery of animal and plant species after the mass extinctions. As noted, the fossil record is in most cases insufficient to add much new information about rapid evolution. Luckily, humans have been conducting their own large-scale studies of evolutionary change in the ongoing domestication of animals. The rate at which dogs have gone from their wild ancestors to the many distinct "breeds" is a case in point.[9] But science is loath to use dogs as experimental animals, so experimentalists recently used chickens as study animals. Chickens were domesticated by humans thousands of years ago.

One of the really amazing transformations in morphology and in other functions seemingly all dictated by the chicken genetic code was recently performed in Sweden.[10] The results were unexpected. The many varieties of domesticated chicken originated from *Gallus gallus*, a tropical bird colloquially known as the red junglefowl. The first record of humans keeping them, presumably penning them and breeding them as food, comes from around 8,000 years ago. Darwin believed changes were the result of humans imposing their own dreadful kind of natural selection in forcing evolutionary change, but that the changes that were saved came from the slow process of mutation. A breeder might notice that one of his chickens was more plump than all the rest, and then he could breed that chicken with another plump chicken. But how did the new trait of plumpness come into being in the first place? Being plump would require far more than a single genetic change. A fat chicken would need commensurate redesigns of muscular systems, blood vessels, dynamics of growth, and on and on. It turns out that domesticated chickens are fatter—twice as fat, in fact, than the jungle-fowl they were evolved from by humans. Did all of the changes take place by Darwinian mechanisms? Darwin used examples of domestication to support his arguments that the formation of species was caused by natural selection. But one aspect that Darwin remained silent about was behavior. Domesticated animals are strikingly different from their original ancestors in many aspects of behavior.

The introduction of one paper[11] on the evolution of domesticated chickens lists the ways that domesticated chickens differ from *Gallus gallus*: They grow faster, become sexually mature at a younger age, lay more and larger eggs, show a wide variation in plumage color and structure, and have a different set of behaviors compared to the nondomesticated root stock of modern-day chickens. These newly evolved chickens appear to form fewer relationships with other chickens and have fewer social interactions. They are far less aggressive not only toward competitive chickens (for mates or food) but even toward potential predators.

Chicken domestication seems to have occurred at different times in different places, and the extensive geographic separation strongly suggests that there should be multiple cases in the past of domestication 7,000 to 8,000 years ago in China, but what appear to be chicken bones of domesticated forms do not appear in the more western parts of the vast Asian continent until 4,000 years ago, when they existed in the Fertile Crescent of the Indus River Valley. From that point and time, however, domesticated chickens spread quickly into Europe and northern Africa.

What struck the Swedish team of Daniel Nätt and his colleagues was that the sum of differences seemed greater than what the difference in the genomes of domesticated chickens and *Gallus gallus* would seem to warrant. How did such similar sets of genes produce so much change in subsequent generations in such little time? Thus, the team set out to compare the degree to which the epigenome, the sum of the domesticated chicken genome with its methylated sites on its DNA, might differ from the genome of *Gallus gallus*.

The same team had earlier shown that inducing extreme stress caused epigenetic changes in the brains of domesticated chickens. Domesticated chickens show extensive methylation of their DNA compared to the wild birds. These patterns of methylation were shown to be heritable and extended to gene expressions far beyond the effects of stress (which induces stress molecule states that are heritable). The conclusion was that variation in the domesticated chickens was greatly increased in many generations by the nature of the epigenome, which produced a wide variety of different epigenetic states affecting characteristics ranging from behavior to physiology. To cap it, the experimenters then crossbred the highly methylated domestic chickens with wild birds and found that the offspring showed the methylated states, and continued to do so for eight generations.

The story of chicken domestication is useful in visualizing the kind of evolutionary change that followed in the wake of mass extinctions. A new environment of wholesale change and difference: loss of predators, and new food sources. Both created new kinds of chicken behavior. And soon after, new kinds of chickens. So too was the world changed for the survivors of mass extinctions. To the many kinds of tiny mammals surviving the dinosaur extinction at the end of the Cretaceous, there was food aplenty if one could eat rotting dinosaurs, and no longer were there fast dinosaur predators everywhere. Mammals could live in the daytime and stop being nocturnal. Food and no predators! And rapid evolution through epigenetic processes.

While the scientific study about mass extinctions certainly looks at both the causes as well as the ultimate body count, there is another aspect of these relatively rare but life-changing events. The extinction of so many species in short periods of time initially leaves the planet with far less life than it had prior to the extinction. An emptied world is a world of opportunity, and each of the biggest mass extinctions was followed by the formation of not only new species but often entirely new body plans. The animals and plants of the early Mesozoic, some millions of years after the Permian extinction, looked far different from the pre-extinction creatures. Also puzzling is the rate at which these new kinds of species appear in the fossil record. Their evolution was *fast*.

AFTER MASS EXTINCTIONS

The concept of "extinction debt" says that while the proximal cause of a major extinction killed many species quickly, other species hung on in ever-smaller numbers. These species were doomed as well, but they just did not die off as fast as the others, and many were still around as newer species evolved in response to the first flush of extinction. In ancient extinctions, such "recovery fauna" was mixed for some thousands of years with the slowly disappearing species from before the crisis. This same argument can be made for many—perhaps too many—species on Earth now that are rapidly dwindling in number: elephants, giraffes, tigers, and on and on. The difference is that we have zoos—but if a species survives only in a zoo, isn't the species essentially extinct? Does it still exist?

During past mass extinctions, the wholesale killings produced environments so different that they triggered the formation of the recovery fauna, and did so quickly. In this sense, the domesticated chickens are a kind of recovery fauna. Mass extinctions were a double environmental whammy: first the killing phase (whether caused by the environment being suddenly too hot, too cold, too poisonous), then the second phase (no food, no mates, no symbiosis) caused disappearance of most life. Such radical environmental change can certainly be thought of as producing that *other* phase of evolutionary change: epigenetic change. The slow change by Darwinian means would not be fast enough to promote species survival.

After every one of the major past extinctions, there were not only new species produced but often new *kinds* of species. For instance, the famous dinosaur-killing K-T mass extinction was not followed by another evolution of the dinosaur body plans—except for birds, which were far smaller than the average dinosaur. Instead, there was a wholesale appearance of the many new body plans produced by the surviving mammals. One possible reason for this is that the world after the great impact was environmentally different and did not return to the same conditions that had produced and favored the dinosaur body plans.

In a previous book,[12] Alexis Rockman and I posited that the extinction of the world's megafauna was the opening blow of a current mass extinction, now under way for 40,000 years, begun by the elimination of the Australian marsupials as well as the giant lizards that lived there. This is certainly how the past mass extinctions went: big animals went first.

The "new" recovery fauna of the late Pleistocene into the Holocene (which combined could be redefined as the Anthropocene, beginning 40,000 years ago) was quite different in anatomy, and perhaps behavior, from the dominant species that were going extinct, including most large mammals that had not been domesticated. Just as the formation of new kinds of organisms through the process of domestication is now known to have triggered epigenetic pathways that produced a cornucopia of new shapes and behaviors of dogs, horses, house cats, and chickens, so might the new recovery fauna of the Anthropocene have produced some surprises. Who would have foreseen the domestic turkey? Or the beauty of tea roses or flowering dogwoods? Or of English bulldogs? Or the races of humans, for it can be argued that current humanity became domesticated with the extirpation of the globe's human-eating carnivores (big cats, most wolves, most larger bears) combined with a new constant food supply (agriculture), both beginning about 10,000 years ago, which also marks the start of a major increase in human genetic change, as will be expanded on below.

As noted earlier in this chapter, the process of domestication has provided some of the best examples of evolutionary change through epigenetics, and in some ways, the "modern" efforts at biological engineering of food, animals, and plants are but an extension of the earlier kinds of domestication (but using radically different methods, such as implanting new genes). Until the end of the twentieth century and advent of the twenty-first, the natural world had *never* evolved a square tomato, or any of the numerous other genetically altered plants and animals now quite common in agricultural fields and scientific laboratories. Just as physicists are bringing *un*natural elements into existence in the natural world through technological processes, so too has our species invented new ways of bringing forth varieties of plants and animals that would never have graced the planet without human intervention. The new genes created and spliced into existing organisms to create new varieties of life will have a very long half-life. Some may exist until life is ultimately snuffed out by an expanding sun some billions of years in the future.

Humans have profoundly altered the biotic makeup of Earth. We have done it in ways both subtle and blunt, yet in so doing we may have not only changed the organic world but tipped that world into a state of evolutionary change that is dominated more by epigenetics than by Darwinism.

Nature is composed of ecosystems that can roughly be identified by how energy flows through their variety of organisms, all adapted to specific environments. One of the earliest recognized aspects is that the number of organisms eating other organisms can be subdivided by what they eat and "who" they are eaten by. On any terrestrial grassland, for instance, there is a great deal more biomass of grass than grazers (nowadays cows, mainly). And, in similar fashion, the mass of grazer predators is far less: the medium-sized wild cats, wolves, and smaller bears are far rarer. Atop this "pyramid" are the top carnivores, the biggest and thus the rarest. During mass extinctions, the animals that were fewest in number and the ones that depended on a large biomass of food were the first to go extinct. This is because mass extinctions killed off species by killing individuals. The larger the population, and the more wide-spread they were, the harder they were to kill.

AFTER THE PERMIAN MASS EXTINCTION

One of the most famous of all fossil field areas is the Great Karoo Desert of South Africa. A dry, dusty landscape used now only for sheep farming, the Karoo holds the world's most abundant and complete record of the crucial time interval ending the Paleozoic era and starting the Mesozoic. There are more and more kinds of Permian- through Triassic-age verte-brates in the Karoo than anywhere else on Earth. Because of that, it should be the best site on Earth to measure the rate of evolutionary change in the immediate aftermath of the Permian extinction.

The environmental cause of the Permian extinction, as well as several others known as "greenhouse extinctions," was rapid global warming produced by the release of massive amounts of greenhouse gases from enormous volcanic activity. Great meandering rivers approaching the size of the Mississippi, Columbia, or Nile crossed the southern portion of the African continent some 252 million years ago. Rivers produce quite recognizable strata as they migrate across a flood-plain. Because rivers are rarely straight in shape, any curve has a higher velocity side and lower velocity side, based on water depth. Erosion digs deeper into the bank in the high-velocity part of the curve, while the opposite side, called the "point bar," gradually fills with softer sediments. But one of the surprises of research into these beds at the end of the twentieth century was coincident with the mass extinction: the rivers changed morphology, from the meandering kind to what are known as braided rivers—many smaller anastomosing streams filled the river valleys where one large river used to be. And this change—a major environmental change for the animals living through it—required new kinds of adaptations.

One of the most amazing aspects of this transition comes from the paleontological record of these beds. While the Karoo beds, as noted above, are indeed fossiliferous, this a very relative concept. The skeletons of larger vertebrate animals that died fell into the rivers and their remains were swept some distance before coming to rest. Their skeletons entered the rock record by burial from river-borne sediments, yet there they are never as common as invertebrate fossils. Today, there are untold millions of deer, sheep, and cows living near rivers. The sheep and cows are minded by people to some extent, but not so the deer. In eastern Pennsylvania and rural New Jersey, for instance, there are so many deer that they are a major hazard to drivers at night. Yet, on a given day, you could walk the length of any larger river in these states and rarely if ever find a deer carcass snagged and unmoving at a place on a riverbank that would seem to allow burial by sediments in the near future. Those that are present are soon disarticulated, the bones scattered and carried off by various scavengers.

Such was also the case in the late Permian beds of the Karoo. Over the most recent time interval, known as the *Dicynodon* zone, finding a skeleton in place takes a great deal of searching. A five-year project of collaboration between paleontologists of the South African Museum and the University of Washington ended up showing that the discovery of each identifiable fossil of the *Dicynodon* zone took an average of eight hours of searching by each experienced paleontologists.[13]

But given enough time and people, the research by this project and by earlier generations of paleontologists has yielded a spectacular diversity of mammal-like reptiles, including perhaps a dozen different kinds of herbivores, from small to the largest, *Dicynodon*, itself a cow-sized reptile. Perhaps half that number of carnivores have been found among the assemblage; the largest of the carnivores were lions and bear-sized gorgonopsians, or gorgons. This community also had a spectrum of smaller carnivores belonging to different genera.

The North American communities of herbivores and carnivores prior to the 1800s can be used as an analogy in terms of the various "trophic" groups having similarities to those reconstructed in the Karoo Permian. In North America, there were (and still are in parks) a large diversity of rodents and many kinds of herbivorous animals. Various species of deer ranged in size from small to large, and were themselves dwarfed by elk and moose. The predators of herbivores ranged from skunks and weasels at the small end to bobcats, lynx, cougars, and black bears. There were also wolves and the largest of the carnivores, the grizzly bears. There were many kinds of herbivores and many kinds of carnivores, but the populations of herbivores far outnumbered those of carnivores.

In both of these wild communities, selection pressure was intense. The herbivores spent every day of their lives using sensory adaptations to detect approaching carnivores, and they used locomotion adaptations to escape during predatory attacks. They also needed very specific adaptations to deal with seasonal vegetation changes across their habitat ranges.

Selection pressure was also intense for the predators. Starvation and death was the result of being poorly adapted. Carnivores needed keen senses to find prey, mobility to chase down prey, and motor skills to kill prey. These called for three very different kinds of anatomical and physiological adaptations.

Yet as one walks upward through time, across the extinction boundary, one can walk over bed after bed of thin, variegated mudstones and immediately enter tens of meters of thick-bedded bright red strata. In these blocky red beds are uncounted remains of a single fossil—the sheep-sized survivor of the extinction, a herbivore known as *Lystrosaurus*. The sheer immensity of the numbers of *Lystrosaurus* is staggering. The beds have yielded the rare carnivore at best. There is only one, the forerunner of crocodiles and alligators, that would have had the size and the anatomical weaponry sufficient to kill *Lystrosaurus*, and it only appears thousands of years after *Lystrosaurus* appeared.

The reptilian carnivores may not even have lived among *Lystrosaurus*, let alone eaten them. They are associated with ponds and shallow rivers and may have been semiaquatic. The other carnivores that are found among *Lystrosaurus* were all tiny compared to the abundant herbivores of this zone. Most seemed to have lived in burrows in the earliest Triassic, as *Lystrosaurus* did too, based on the presence of burrow-shaped discontinuities in the mudstone beds.

The bad news for evolutionary biology is that the die-off of the Permian animals found now as fossils in the Karoo happened over a thin interval of strata, and thus a relatively short time. But the bigger surprise is that the new fauna—of the abundant *Lystrosaurus* and the few carnivores among them—was instantaneous. These creatures appeared suddenly, as if herds of sheep had been dumped onto the world, with a few house cat–sized carnivores scurrying among the legs of *Lystrosaurus*, trying not to get trampled while *Thrinaxodon*, a tiny insect-eating mammal no bigger than a mouse that is ancestor to us humans, searched for beetles. For both, it was a new world. In our world, small carnivores are eaten by larger carnivores, but in the Karoo world, both *Lystrosaurus* and the carnivores got a free pass. There was no selective pressure at all. *Lystrosaurus* had no need to detect carnivores coming, or to be able to run fast and escape them, because there were no large carnivores.

In this evolutionary experiment, the variety of those *Lystrosaurus* lucky enough to have survived the great extinction ran wild. The fossil record of the Great Karoo Desert shows that *Lystrosaurus* acted just like the animals that humans have domesticated. Prior to the Permian extinction hitting, this land-based community of large terrestrial animals of many species and shapes grazed on the rich, low, leafy vegetation on land and in shallow ponds. These were stalked by large predators, most notably the "lion-lizard" hybrids known as gorgons. But none of these large carnivores—or even the medium-sized carnivores—survived the mass extinction. The herbivores that did quickly increased in number.

It was far faster for the surviving *Lystrosaurus* to restock a world than it was for a new predator to evolve from scratch, or for one to evolve from small, insect-eating protomammals. Thus, for hundreds of thousands of years following the mass extinction, the earliest Triassic-age Karoo ecosystems were inundated by the lamb-sized (and presumably lamb-tempered) *Lystrosaurus*. After the extinction, they took a drastic reduction in average size. According to Cope's rule, animals should increase in size as they evolve. However, when predators are absent, things change. Birds lose their wings, as it is no longer necessary to fly. And without predators, herbivores got smaller. It took a long time for the carnivores to evolve to be large enough to take down even the new Triassic-age mini-*Lystrosaurus*.

Something similar occurred when we domesticated animals, like chickens. We removed their predators and the constraints of natural selection caused by the need to run fast or hide or fight back disappeared. Anatomical and behavioral characteristics changed. Removing predators is one of the most significant of all environmental changes. According to Lamarck, environmental change leads to behavioral change, which leads to anatomical change. Soon after the mass extinction of the Permian, an enormous variety of new shapes appeared among the survivors. They did so in the same fashion as the domesticated chickens did—through rapid, heritable epigenetic change.

Our own species is also subject to natural laws of evolution, which include the epigenetic ones. If we as a species underwent conditions analogous to those that domesticated, caged chickens adapted to when we killed off our predators, and when we changed from diets of scarcity and switched to agriculture, then there was little need to defend against predators or to spend most of our days hunting for food. But there was probably coevolution taking place as well. We created dogs and chickens and cows and wheat and cereals. And they created us. We "modern," post-agriculture humans—are we domesticated humans? We make fun of cows and chickens and sheep as stupid vassals. But how do they view us? If past is prelude, our own domestication should have coincided or soon after witnessed an outpouring of different human races.

With our domestication (or not), what is certain is that our habits changed, and then our anatomy, physiology, but probably behavior first of all à la Lamarck. The many kinds of human races can be analogized to the many kinds of dogs, chickens, and other domesticated animals—all the same species, all capable of producing live offspring, but all anatomically different.

Our behavior certainly changed. No longer constrained by natural selection, we began living in large groups (with the concomitant disease and parasites) and producing new foods. Unbound, we radiated out into large swaths of land. We have set fire to entire continents, and thus have initiated the kind of extraordinary environmental change predicted by Lamarck as being necessary to produce the greatest organic change through time, the true definition of *evolution*. Clearing the land in North America, New Zealand, and Australia caused the evolutionary change of the flora to one of fire-resistant plants in landscapes where such species had existed only in small numbers prior to the arrival or evolution of fire-branding humans. How did the intricate and complicated morphological, physiological, and even behavioral mechanisms necessary to change to a plant species able to survive a yearly burning develop quickly enough to avoid extinction? Epigenetics, in all probability.

We have wiped out entire species and decimated countless more either to suit our needs for food or security or simply as an accidental by-product of changing the landscape to favor our new agricultural endeavors. We have changed the role of natural selection by favoring some species that could never otherwise survive in a cruel Darwinian world over others of estimably greater fitness. We have created new types of organisms, first with animal and plant husbandry and later with sophisticated manipulation and splicing of the genetic codes of various organisms of interest to us. The presence of humanity has begun a radical revision of the diversity of life on Earth—not just the number of species present and their abundance relative to one another, but the evolution of organisms with genomes that are not the product of natural selection.

CHAPTER IX

The Best and Worst of Times in Human History

OUR species is one of the newest in the history of life. Our exis-tence is not measured in millions of years, or even half millions. A fascinating experience would be to somehow bring a member of the very first *Homo sapiens* back to life to compare to us now. The closest we can come is by comparing skeletons, but we cannot know many of the details of the genome of our most ancient members because there is no DNA from human bones of the first of our species. Compared to other mammals, we are young. Although hominids—the group we belong to—are not.

Paleoanthropologists have done a remarkable job in deciphering the where and when of the speciation event that produced humans.[1] The human family, called the Hominidae, seems to have begun as many as 3 million to 4 million years ago with the appearance of a small proto-human called *Australopithecus afarensis*. Since then, our family has had as many as nine species, although there is ongoing debate about this number as new discoveries are made and new interpretations of past bones make their way into print. But the most important descen-dant of the early, pre-Pleistocene hominids is the first member of our genus, *Homo*, a species named *Homo habilis* ("handyman") for its ability to use tools. It lived about 2.5 million years ago and gave rise to *Homo erectus* about 1.5 million years ago, and *H. erectus* either gave rise to our species, *Homo sapiens*, directly about 200,000 years ago, or through an evolutionary intermediate known as *Homo heidelbergensis*. Our species has been further subdivided into a number of separate varieties by some, but as coexisting with other members of the genus *Homo*, including *Homo neanderthalensis,* and a poorly studied group known as Denisovans.[2]

Each formation of new human species occurred when a small group of hominids somehow became separated from a larger population for many generations. In the 1960s and '70s, there was a view that modern humans came about from what has been called a "candelabra" pattern of evolution[3]—that all over the planet, separate stocks of archaic hominids such as *H. erectus* all evolved into *H. sapiens* at different times and places. This notion has been debunked. The major evolutionary changes leading to the humans of today began among some small, isolated group of humans.

Few topics are more relevant to understanding how humans have arrived in the present day than those that look for major changes in human biology as well as cultural history. Historians of *human* history essentially divide into three camps. The first group studies past humanity and its cultures from the point of view of fossils, the second uses archeological techniques (excavation of artifacts), and the third studies any kind of written texts or oral histories. The history that is now taught is a combination of these.

Virtually unstudied, however, is whether major environmental changes in human history caused concomitant evolutionary changes not just in morphology and physiology but in heritable human behavior. Relative to that are the clues coming from human fossils and artifacts, but also the ability to approximate the age of new or modified human genes. Among these are investigations into the when and why of significant changes in how the human brain has functioned since the first evolution of *H. sapiens.*

One such event has been called the "cognitive revolution,"[4] thought by its supporters to have begun some 70,000 years ago. According to this theory, the cognitive revolution was a turning point that produced the first real culture, which then served as a second means of evolution through the eventual rise of written language, of public art, and of great advancements in tool use of all kinds. Yet much evidence suggests that this event occurred 70,000 years ago. To its supporters, the cognitive revolution hypothesis was supported by a marked change in culture, itself perhaps triggered by changes in how humans actually perceived the world and communicated about it. It is likely that real genetic changes were involved, and they were significant in determining changes in human intelligence—how the brain and consciousness worked. Unfortunately, fossil human skulls give little insight.

Major evolutionary changes do not take place simultaneously across an entire large population of a given species. For examples, the spread of the genes that make humans lactose tolerant enabled us to use a readily available and nutritious new food source. These novel and competitively superior organisms begin in some small, isolated populations that—if truly superior in their survivability—then spread out geographically and either outcompete those without the gene (often wiping them out in the process) or pass on the new genes to subsequent populations.

The appearance 70,000 years ago of a series of what archeology suggests must have been significant behavioral changes[5] in our species may have originated in just this way: in a small population of humans undergoing genetic change that then spread out across the planet. There are lines of evidence that a major environmental catastrophe affecting a population of our species then living in northern Africa was roughly coincidental with the onset of what appear to have been cognitive changes.

Since humans first evolved, attaining sufficient food may have been their most urgent task. Fluctuation of food supply would thus have had an effect on population sizes of what may have been bands of hunter-gatherers. This was long before the advent of agriculture, and thus the availability or meat and food supplies coming from plant sources would have been dictated in large part by climate. It so happens that the greatest volcanic event in human history is thought to have occurred about 70,000 years ago: It is called the Toba eruption.[6]

In the nineteenth century, the Krakatoa eruption in what is now Indonesia put so much ash into the atmosphere that it caused a massive reduction of photosynthetic plant productivity in many parts of the globe for one to two years. Yet Krakatoa was nowhere near as powerful as the Toba eruption, which may have caused multiple years of near darkness that surely would have caused plant death and thus animal famine as starvation from lack of plants trickled up the food chains to nonherbivores (although contrary studies exist as well).

The Toba eruption in what is now Sumatra has been repeatedly singled out as a major cause of extinction of the *H. sapiens* living in some areas at approximately the same latitude.[7] The size of the explosion and the amount of rock that was vaporized would have caused meters of ash to fall and blanket the area and downwind locations all around the planet. The amount of fine ash would have changed the albedo (amount of radiation or light reflected off rather than absorbed by a surface) of Earth for an unknown period. Sudden albedo change causes atmospheric circulation changes. Not only are plants killed, but the habits and populations of game that *H. sapiens* relied on would have been drastically reduced.

Under such conditions it seems highly probable that the factors that can lead to major evolutionary change indeed took place—factors like reduction of overall population size and isolation of small breeding populations from the larger gene pool. What makes this particular event most interesting is that geneticists have traced current humanity back to an African population living in the northern part of the continent in just such a region, one that became highly perturbed by major environmental change.

The Toba eruption aftereffects would have drastically altered the climate, producing colder temperatures, perhaps changes in rainfall, and for some period a loss of warmth and normal sunlight. From this came a smaller and perhaps quite different kind of food supply, but also possible physical isolation of humans living downwind from the volcano, thus making what had been one large region now something akin to islands in the few places still with vegetation and the ability to support humans.

A volcano explodes and the human population in the area plummets in size. The human gene pools radically change from a bottleneck effect. At the same time, the planet is experiencing one of the fastest recorded global temperature changes. And soon after, the single population that would conquer the world and replace all other varieties of humanity sets off on foot out of Africa, out of the shadow of the most consequential volcano in the world at the time, where temperatures are plummeting anyway.

Among the survivors, some small group underwent behavioral change that became heritable, producing cultural change that is difficult to overstate in terms of significance. Members carrying the new genes then set forth. But was it only a new set of genomes? What is the possibility that the original changes were to the epigenome, not the genome itself? Could the humanity that survived have changed as a consequence of the explosion, as a result of leaving a landscape of resource unpredictability? In the mass extinctions, the first to go were always the top carnivores, as these meat eaters in any normal mammalian ecosystem depend upon meat volume among prey species that is at least an order of magnitude larger than the mass of the carnivores themselves.

The wildest supposition is that because of the Toba eruption, the *H. sapiens* in the northern African region were at first under enormous selection pressure, with most dying off, probably from starvation. But the survivors, mating among themselves, then would have been under far less stress. Fewer mouths to feed, and probably a reduction of human predators as well. Lamarck told us: First environmental change, which leads to behavioral differences, which leads to phenotypic change.

COGNITIVE REVOLUTION II

A second interval of what appears to have been another shift in cognitive human characteristics is thought to have taken place around 45,000 years ago, and here too there was coincidental, major environmental climate change occurring, this time caused not by fire but by ice.

The so-called ice ages are temporally encompassed by the 2-million-year-long Pleistocene epoch (a formal geologic time unit). Throughout the Pleistocene there were successive advances and retreats of continental glaciers as well as radical changes in prevailing climate over much of Earth, causing very rapid swings in seasonal temperature and rainfall patterns in many parts of the globe. There were century-by-century changes in sea level at unprecedented rates.

The most visible evidence of the cognitive revolution 45,000 years ago (CRII) is that it appears to have brought into existence an entirely new kind of human expression, the sublime and beautiful cavern and cave paintings found in many parts of the world.[8] While it may be that similar art was being made for long periods of time prior to CRII, and that it was simply the invention of better, longer-lasting pigments used to paint that caused what seems to be a "sudden" appearance of art, most anthropologists lean toward the interpretation that something novel in human expression appeared. Yet if there is agreement that the visible art appearing with increasing frequency after about 40,000 years ago was novel, there are still questions about why this happened.

The cause of CRII has the preponderance of anthropologists insisting it was "cultural" rather than biological.

CULTURE OR EVOLUTIONARY CHANGE— OR BOTH?

This dichotomy has been recently addressed by the longtime University of Chicago (now Stanford University) anthropologist Richard

Klein, who unapologetically argues for biology as the cause—that an actual evolutionary change took place that caused cultural change,[9] rather than the reverse of that. Although most anthropology departments disagree with Klein, a recent summary beautifully encapsulated his research. It begins with the discovery of what appears to be the oldest human jewelry, physical objects to make ourselves more attractive to others.

The fragile beads, crafted around 40,000 years ago, hail from a Kenyan site called Enkapune Ya Muto, or Twilight Cave.[10] But some anthropologists think they are much more. The people of Twilight Cave may have exchanged them as ritual gifts or tokens—making the Cheerio-like objects the oldest known examples of symbolism. If the beads were among humanity's first symbols, says Klein, they represent one of the most important revolutions in our species' career: the dawning of modern human behavior. As is increasingly well known, many aspects of animal behavior, including in humans, has a genetic basis.

Klein thinks a fortuitous genetic mutation may have somehow reorganized the brain around 45,000 years ago, boosting the capacity to innovate. There is also speculation that not only was the new art related to genetic novelty, but spoken language also radically increased in sophistication.

There was certainly no more time of extended yet oscillatory rapid climate change in the past 200 million years than the Ice Age. While the major perturbations following the dinosaur-killing asteroid of 65 million years ago indeed rapidly changed climate, within a handful of millennia Earth destabilized: The nuclear winter–like effects had passed. But for nearly 2.5 million years—the last 2.5 million years—there was a major advance and retreat of continental ice over the continents and over large areas of now open-water oceans and seas as well. Within that cauldron of change there was a spectacular evolutionary appearance of some of the largest mammals ever seen on Earth, including giant mastodons, mammoths, Irish elk, and giant sloths, to name a few.

Most of these animals can be seen as a response to an Earth (except for Antarctica) moving from a virtually ice-free planet 2.5 million years ago, the end of the Pliocene epoch, to one that was heavily glaciated. A world with a wide tropic and even wider temperate grasslands that remained unchanged from millions of years switched to a world where both ocean and air currents could change global temperatures by many degrees in two decades, as now seen by the many studies using ice cores.

The recentness of the Pleistocene gives a major advantage to paleobiologists trying to interpret the biology of deep-past animals and plants: In some cases, bones and other organic material can yield recoverable fragments of DNA. Especially from the past 50,000 years, these DNA discoveries have shown how remarkably important epigenetic changes were in major evolutionary changes. Most provocative has been the discovery of methylated, epigenetically changed *Hox* gene sites in humans both extinct and extant.[11] *Hox* genes are crucial in the development of any animal, from the simplest to the most complex. These are genetic switches that tell the various parts of a body how to form, when to form, and what shape to form.

Even slight delays or advances when specific *Hox* genes are activated can change the shape of a human skull, causing it to produce, for example, a far more prominent brow. Or increase the size of the nose. Or cause the chin to recede and the oral area to protrude. Or make the shape of the limb bones slightly different. Everything needed to make two different humans: being almost identical gene for gene but looking remarkable different. Even with the same genes, a different epigenome can make one human look "modern" and the other a prototypical caveman, based, of course, on the famous Neanderthals and the less famous Denisovans.

There are fewer than one hundred proteins that differ between *H. sapiens* and *H. neanderthalis*,[12] spread over the products of 25,000 separate genes. But when the epigenome is compared, there are more than 2,000 differentially methylated sites along the two sets of DNA of these "cavemen" and "modern" humans. Thus, it is not the slightly different amino acid constitution of these separate proteins that makes the profound difference in phenotype. Those are caused by differences in epigenetic factors between the two species, and none more important than differential methylation patterns in the *H. sapiens* and *H. neanderthalis Hox* gene complexes, as a 2014 study demonstrated.[13]

Another major difference was discovered in this study. Gene sites that are associated with various diseases were twice as likely to be methylated than were the same genes in *H. sapiens*. The conclusion is that the Neanderthals, and their relatives, the Denisovans, may have had a different outcome or mortality rate from these shared diseases based on the difference in whether they were methylated or not.

The finding that the *Hox* gene complexes showed differences in methylation patterns might be a vital clue to understanding how rapid and major changes in limb and limb digits (our fingers and toes) took place. The most famous of such rapid changes comes from the evolution of horses, going from small with toes in the Eocene of 55 million years ago to large with a single hoof today. But it also demonstrates that epigenetic processes were at work on humans in the past.

FROM NEOLITHIC TO AGRICULTURE

It is useful to look at what is about to overtake us, as by 2100 the human population will perhaps crest at 11 billion,[14] as a continuation of trends that began 10,000 years ago. Perhaps the most significant change in our biological history was caused by a change in our social history. With the invention of agriculture came cities, and with cities came crowding, new diseases, and ever more people. Rather than the evolutionary history of our species slowing, the change to cities and agricultural fields where once there were only small bands of hunters and

gatherers has caused our evolution to speed up. It is most scientifically
parsimonious that heritable epigenetic effects were a big part of this
history. It certainly is among the animals that preagricultural humans
hunted during the last glacial interval, and we know this from direct
measurements of DNA from Ice Age mammals and humans that show
epigenetic methylation sites.

The conceit of Michael Crichton's *Jurassic Park* novels, and their
accompanying movies, was that DNA can remain unchanged over
millions of years if preserved in amber. "Unchanged" is a tall order
for any biomolecule over years, let alone more than 200 million years.
Yet a cadre of investigators being called "paleophysiologists" is
collecting DNA from the Pleistocene ice ages. They are finding not only
that important information can be gleaned from ancient beasts and
ancient humans alike from that time but that some epigenetic marks
are being preserved.

One of the great stories coming from late twentieth-century paleobiology concerned the cause of the extinction of North American
mastodons at the end of the North American Ice Age, some 12,000 years
ago. University of Michigan paleobiologist Daniel Fisher used detailed
measurements from mastodon tusks[15] to determine whether the final
representatives of what had been a very long-lived species were under
stress from climate change (with loss of food, perhaps) or hunting—
presumably overhunting from humans, based on finds of mastodon
skeletons with spearpoints still within. Fisher knew from the study of
modern elephants that the female reabsorbs parts of her tusk when
pregnant so that more calcium can go to her growing fetus. Fisher also
knew that elephants produce more young when they are being hunted
down by humans in the present day, so as to make up numbers. All of
this is known from tusks.

Now it is also actual, readable DNA, still organic, that is giving
information about the levels of stress in ancient Ice Age mammals, and

as is the case with the discovery of methylation in ancient human DNA, these studies are documenting epigenetically driven changes in levels of stress in ancient animals during the last phases of the Ice Age. In the forefront of this is Alan Cooper of the University of Adelaide, who spends parts of every summer hosing down glacial deposits in the Arctic, looking for fresh, 20,000-year-old bones that can yield DNA. Epigenetics is squarely in the wheelhouse of his studies.[16]

Cooper has shown that the level of epigenetic markings increased when new regions of the globe were invaded by humans. His work has shown that methylation levels increased radically when Ice Age bison and musk ox from the Arctic regions of some 20,000 years ago first encountered humans migrating across Asia, one of a succession of human migrations that would bring humanity in numbers to North America.

As late as 25,000 years ago, our far-flung species had made its way to every continent save Antarctica. Yet, in spite of being world girdling, we remained at low population numbers; there may have been fewer than a million humans in total. Hunting and gathering is not the sort of lifestyle that requires crowds—just the opposite. Small bands of hunter-gatherers did much better than larger groups in areas where game was prone to become scarce due to the efficiency of these smarter humans. There was no worry about fouling water from human waste— there were never many humans at any one spot, and there were no permanent settlements of any size at all. But human population eventually outgrew the number of large animals necessary to sustain it, and by some 10,000 years ago they invented agriculture.

Yet, with agriculture, the rules of human survival and lifestyle radically changed. Now large numbers of humans could be sustained by this predictable food source, one that was very labor intensive, so humans needed permanent settlements to watch the crops grow. Child mortality dropped; with permanent settlements, predation on humans dropped as the local predators were necessarily wiped out. But at the same time, sewage and sanitation brought new dangers. With new food sources came digestive problems. With the crowding came new diseases and new parasites. Each of these put the gears of human evolution into higher speed. But some of these clearly had epigenetic side effects, only now being discovered, some of which were devastating.

Epigenetic change takes place when a human encounters a significant environmental change. Being among many more humans than previously within a band of hunters was such a change. Being inundated with the hosts of new microbes in our guts coming from new food and fouling of food and water from the fecal material of those early crowded cities also unleashed epigenetic change.

Our species is now affected by a greater diversity of factors that can trigger epigenetic change than at any time in our species' history. The most important of these are the chemicals we surround ourselves with, the food we eat (or do not eat, during famines), and the diseases we acquire during our lives from microbial invasion or even through inheritances.

In the past two decades, substantial research attention has been given to the structural differences that emerge when the brain is exposed to divergent sensory experience. This research must be somehow related back to, and seen in the context of, the greatest of all social changes, the nature of human societies pre- and post-agriculture. If agriculture was the linchpin that allowed city states to raise and maintain professional armies paid in food and territory, it really behooves science to look at how these times affected not only our cognition but more importantly our behavior, and how malleable our brain is to change, and how it was able to change from a hunting-gathering Ice Age lifestyle to an agricultural one.

The changeability of the brain is evident because it can be demonstrated that external conditions can transform the course of its development. This phenomenon, referred to as neural plasticity,[17] is especially potent when exposure occurs at certain vital times. These sensitive, or critical, periods are the windows of opportunity when environmental influences have their principal bearing.

In order to understand the momentous character of this intervention, it is necessary to have a rudimentary understanding of how the nervous system is constructed and how it operates. An infant is born equipped with all the neurons that she or he will ever have in the cerebral cortex. Yet, because of the way these cells connect, there are infinite possibilities in the way neural networks can form. The consequent whole, the physical brain, with all its seemingly infinite ganglion connections, is created and re-created through this process. While the macroscopic features of the brain are, of course, essential, it is in the formation of the neural networks where plasticity lies. Brains immersed in computer use, texting, gaming, and unending communication become wired differently than those of previous generations. Until recently, the response has been societal unease but classical Darwinian academic disinterest, since none of these changes could be inherited. Neo-Lamarckism totally changes the playing field. Real physical changes are taking place in our brains. And real epigenetic consequences await us. In some sense, we remain Ice Age relics newly immersed in an alien computer age.

POST-AGRICULTURE HUMAN HISTORY EVENTS THAT MAY HAVE CAUSED EVOLUTIONARY CHANGE

Any online search asking for a "brief history of human civilization" is more often than not responded to with temporal categories such as pre- and post-invention of writing, with the implication that history only began when humans were capable of writing it down for future generations. That turning point is rounded off to 5,500 years BP (before present).

The time line[18] of this prewriting time interval, often called "human prehistory," is broken (in years before present) into Middle Paleolithic, Upper Paleolithic, Mesolithic, Neolithic, Bronze Age, and then the "younger stuff," going by millennia.[19] There are at least hundreds of thousands of pages devoted to the history of humanity. Most of that history is organized by time intervals. But our history can also be sorted by the kinds of events that appear to have been so major that they have come down through time as either written records or oral tradition.

Some of these events are analogous to events in the history of life that produced both extinction and evolutionary change. Human warfare, genocide, famine, and global disease are such analogues. Perhaps each has caused evolutionary change in us no less profound than is known to have occurred by the advent of agriculture and the crowding of humans into cities starting 10,000 years ago. A difference, however, is that unlike a mass extinction, which was global, until the twentieth century there were no "world wars," or global famine or disease that affected all humans across the world at the same time. But certainly entire regions that were affected by catastrophe may have had human populations that were evolutionarily affected.

The choosing of major events in human history is of course subjective. Some, however, are common in most summaries. Most involved short-term changes that in turn changed the environments inhabited by human populations. These include, in no specific order:

1. Food change or first appearance (such as the inception of some new kind of food).
2. The advent of a powerful and historically important new ruler.
3. New building: a historically important construction, such as the pyramids.

4. War: this also included the start or end of a war; thus some wars got two data.
5. Disease: a major plague.
6. Famine: a historically relevant famine, such as during the Black Death.
7. New region: the appearance of humans for the first time in a new region, or the conquering of one already peopled place by another.
8. Natural disaster: major earthquakes, volcanic events, or floods.
9. Religion: This could be the appearance of a new religion or the conquering of a new area and imposing of a religion so that a major conversion took place, such as the birth and territorial increase of Islam or Christianity.

The above summary of human history over the past several thousand years that was rather arbitrarily chosen from many available on the World Wide Web. This particular summary is mainly devoted to Europe.

The authors of this list clearly thought that the ascension of new rulers had pride of place, followed, respectively, by religion and war. Some of these events, such as the conversion of a large population to a new religion, also caused a conquering event for more territory by one people over another, and, inevitably, war.

Each of the categories used would have had major effects on the sum total of relative hormone levels in large swaths of humanity. Some interesting questions emerge. If we could somehow sample enough bones from medieval graves of people who died before and after the Black Death in Europe (but not of the plague), it would be interesting to see whether the European populations showed more epigenetic marks—more methylated sites—before or after the plague, or if the plague actually had little effect.

And from that comes a more interesting question: How much heritable evolution did the above categories cause not only in humans but in the global biota? Certainly, domestication produced large-scale evolutionary change in the affected animals and plants. But the rapid increase in human population over the past thousand years has caused stress to many supposedly unaffected organisms. How much evolution has humanity actually triggered in Earth's biota?

MODERN HUMAN BEHAVIOR—THE STILL UNKNOWN ROLE OF EPIGENETICS IN ITS FORMATION

Let's think like a physicist for a moment and conduct what Einstein called a "thought experiment." In ours, we go back to 2 million years ago to snatch a male and female adult *H. erectus,* to thousand years ago to pluck two ancient Britons, and then grab a pair of present-day Americans. We get a serious big-time psychologist to teach each of the six some quantification of the so-called Big Five personality traits (openness, conscientiousness, extraversion, agreeableness, and neuroticism).[20] Openness includes imagination or intellectual curiosity; conscientiousness deals with carefulness and organizational capability; extraversion is associated with gregariousness and the tendency to seek out stimulation; agreeableness is just as it sounds, but it also deals with a level of cooperation and compassion; and neuroticism is the ugly duckling of the five, dealing with negative emotions and depression (it is also called emotional instability).

Then we bring in an evolutionist and ask: Which, if any, of these traits might have best served natural selection for these humans during the era in which they lived?

What was new was the first scientific attempt to quantify the degree of heritability of these traits. The surprise was that only two, openness and neuroticism, showed relatively high levels of heritability.

One wonders why evolving humans would not have selected for the maximum ability in those traits leading to organization, cooperation and compassion, and gregariousness. All would seemingly be important among small bands of humans trying to survive in the world-climate-gone-mad caused by the radical, Pleistocene-temperature/sea-level/ice-volume climate changes. And most curious: Why would neurotic traits, none of which seem to be of any use in a dangerous world, remain so heritable? As in so much of human biology, no single gene or single trait is ever acting alone, it seems. Something as complex as human behavior is surely more divisible than being describable by only five traits, and within these, genetics is a prime determinant. But the reality is that many of the more observable behaviors of humanity that have negative influences, such as risk-taking (which is often associated with alcoholism and drug addiction as well as crime) and the degree to which a person is susceptible to stress—through the formation of stress and pleasure molecules and how quickly these are expunged from the liquid parts of the human body—are decidedly major determinants of how we act.

There remain skeptics that any aspect of personality is entrained by genes and thus is heritable as well as affected by natural selection. But as in so much else dealing with human biology and its relationship to genetics (and epigenetics), studies on human twins tell a compelling story. And as is so often the case, the first study to actually look at quantitative heritability[21] of the Big Five skipped any discussion of how epigenetics might be an added and overlooked aspect of heritability.

There are certainly skeptics[22] about the role or even existence of personality entrained genes. It remains a hotly contested research area, and a difficult one. Here is a case where rats cannot be used as study subjects if we want to know about complex human behavior.

The most common type of epigenetic change, resulting from methylation, is thought to be a cause of many of the sometimes profound differences between otherwise genetically identical twin siblings. One of the most fascinating results came from a paper by Zachary Kaminsky and colleagues.[23] As in any study of identical twin humans, the weakest link in methodology is always the small number of samples. Of those studies, the results comparing the personalities of a pair of middle-aged twins who had quite different jobs and professional lives show the stark difference that epigenetics can bring.

One of these twins was a war journalist, one of the highest-stress professions; some even consider war journalists to have stress levels akin to those of the soldiers themselves, as the journalists are always trying to be on the front lines. The other twin could not have had a more dissimilar profession, working in an office, in a white-collar clerical position. The war sister reported on violent battles and the most gruesome atrocities from the African continent for two years. She married late, never had children, lived often in many challenging environmental conditions, and saw many colleagues or befriended soldiers die. The other twin married early, had two children, never smoked or drank much, and had a peaceful life. The war twin was methylated to the max: smoking, drinking, experiencing multiple cases of life-threatening situations, and witnessing death in its more horrible forms. And yet, paradoxically, it was she who, through complex psychological testing, showed less propensity for reacting to stress, less inclination toward depression, and a different and lower level of risk-taking. It seems as if the epigenetic changes she underwent made her lifestyle more palatable—that she adapted to the stresses of her job in a permanent, life-changing way.

It is so difficult and invasive to study human behavior. There are privacy concerns, especially now that it has been revealed how social media sites actually sell data about us that can be used to "pigeonhole" us to specific products, sales, or even political candidates. The role of epigenetics in behavior will perhaps remain enigmatic as more and more of us humans refuse to voluntarily or involuntarily serve as lab rats.

CHAPTER X

Epigenetics and Violence

I s there a behavioral heritability associated with two of the most basic human emotions: love and hate? And if there is, would not times of war be among the most influential of all forces driving human evolution? There have been many times of war, yet no time in history (beyond San Francisco, circa 1967, and that one tongue-in-cheek) noted for its peace and love.

It may be that love and hate are different expressions of similar genes, or perhaps the result of different amounts of proteins being called for in different situations. But what is certain is that quantifying the "amount" of hate or love someone feels is still impossible. However, we can gauge their extent by assessing the propensity toward actions that love or hate manifest.

We can ask if violence of humans against other humans within their own societies can be shown to have changed through time. The second and perhaps more profound question is whether violent behavior is passed on. Hate is often lifelong, where violence or the propensity for violence can be and usually is brought forth by short-term stimuli. At the end of a difficult day a driver is cut off in traffic. He honks at the other driver, who flips him off. The confrontation escalates and ends with violence. Short-term violent reaction is all too common among humans. Too often it is caused by the lack of restraint; or short-term decisions made without reflection; or people who are risk-takers, behaviorally erratic, and unpredictable. All of these might be tied into genes that themselves can be changed in a lifetime and then passed on through heritable epigenetics, new studies suggest.

A SHORTER HISTORY OF WAR

War colleges through time have required their nascent warriors to study the history of warfare in ways and means no less diligent than art school students look at the Old Masters. Soldiers and officers need to know not only when, where, and how (troop size, kinds of weapons, etc.) but the most elusive: why. Who knew that one of the most destructive wars in human history, World War I, would be set ablaze by the assassination of an Austro-Hungarian aristocrat? Future officers have to learn how to fight and part of that can come by studying the past. Sources abound, but my own comes from a combination of summaries from the United States Army War College[1] and insights from Tim Hetherington and Sebastian Junger's documentary *Restrepo*.[2]

Has the seemingly perpetual existence of global warfare changed us genetically and/or changed some percentage of soldiers in ways that produce heritable epigenetic change? The U.S. Army War College sources suggest that large-scale war did not come into existence until there were urban societies. Within a millennium of this, humans only equipped with varieties of stone tools began the evolution of more complex metallurgy. By around 5,500 years ago, the nascent agricultural societies clustering around the Mediterranean and into the Fertile Crescent and parts of Asia were using bronze. As metal was fashioned into war chariots, helmets, breastplates, and metal-tipped weapons, warfare rapidly became more strategic and lethal, especially coupled with the invention of cavalry. But did the appearance of weapons increase the prevalence of war? Or did war cause a meaningful change in the human gene pool, not only in the *totality* of genes but in the *percentages* of crucial genes that are associated with defeat, fear, anger, and triumph? Did humans evolve war, or did war help evolve humans?

Certainly the move from small pastoral populations to urban centers powered human evolution. New studies have shown that the rise of agriculture and the introduction of novel kinds of food led to human crowding in an entirely new way, which led to the spread of new diseases and new parasites to populations that previously had not been subjected to either.

The first two agricultural states were Egypt and Sumer. In both cases, the agricultural revolutions came first, followed by new societies that became subject to new kinds of top-down governmental rule. Administration followed, and what had been warrior castes of nomadic societies became the soldier castes of well-administered societies. Soldiers did not have to farm to stay alive—they were fed.

The case has been made that from 6,000 to 4,000 years ago, human history saw some of its most significant social change, and the interesting question is: Was this time too among the most significant in human genetic change?

There is a powerful understanding from biology that evolution drives a warfarelike equivalent between predators and their prey. From the Cambrian explosion of animals, beginning some 540 million years ago, rapid evolution has been driven by visual predators and their visually found prey. This continued its rapid escalation amid the evolution of offensive weaponry (to gather and then eat live or dead prey) and also defensive adaptations (to detect the predators and then to escape from them through movement or stationary defense), including employing armor or defensive weaponry that has an offensive capability, such as the active use of poison chemicals sprayed on the predator. It is naïve to believe that behavior enhancing the predators' ability to get prey and the preys' ability to change behavior to increase survivability were not evolving equally as fast.

Every once in a while, some new adaptation makes for enormous change, with one side gaining great advantage, such as predators developing the ability to crack open shelled mollusks. When new kinds of fish, arthropods, and even snails figured out how to either crack open or drill holes into the armored herbivores, the world went topsy-turvy for a while, until the defenseless creatures produced new defenses, went to new places, or went extinct.

The major proponent of identifying the history of predation within the history of human warfare, Gary Vermeij of the University of California at Davis, has called the Jurassic and the Cretaceous periods the "Mesozoic marine revolution."[3] The evolution of the chariot, the reusable rifle, the war-capable airplane, and atomic weapons all produced similar, if short-term, dominance. These single adaptations also radically changed the entire "DNA" of the armed forces they were a part of. Armies that attacked each other on foot became armies with significant numbers of cavalry and chariots. Armies of riflemen, air forces, and ICBMs also needed humans to build weapons and guard and shoot them. Relative to entire armies, small new weapons radically changed the entire complex structure of militaries.

Regarding epigenetic changes to the gene pool caused by warfare (the effects of combat, the effects of the grieving families, the overall effects of mass death and the necessary economic procurements allowing it), the most radical change would have been from 6,000 to 4,000 years ago. In less than 2,000 years, humans went from warfare that was relatively rare to warfare where not only the size of the combatants approached modern army sizes but the size of *administrative* groups necessary to keep war going (weapon procurement, other supplies, salaries) did as well. And it was in Sumer and Egypt that the world witnessed the emergence of its first armies. Armies are never constructed to only march in parades; with armies, humanity began its love affair with war.

GENES AND VIOLENCE

It remains an evolutionary curiosity that creatures using the most complex biological organ ever evolved—the human brain—still resort to a series of organic molecules coming down through time, through almost a half billion years of time, to produce some of the most "human" of emotions: love, fear, anger, humility, generosity, among so many others. One such chemical that has perhaps had a greater influence on human history is the chemical agent of stress.

The whole of stress, fear, anger, panic, and the urge to flee comes from the cocktail of hormones that include cortisol and the equally powerful adrenaline molecule. The receptor associated with removing chemicals most associated with violence is the glucocorticoid receptor.[4] Someone's fate or behavior might be related to how many glucocorticoid receptors are in a person at any given time. It is one of the basic building blocks of our stress-response system. It's a protein that helps us control the hormones that cause stress: The more of the receptor we have, the better we're able to respond to stressful situations, because while the chemicals will still be created in the fight-or-flight situation, the faster they are scrubbed from the cells and body, the less chance the incident will escalate into violence.

Within the cells themselves, one of the most potent signaling chemicals of all is the enigmatic compound hydrogen sulfide, a potent toxin even at small concentrations. It has been shown to be one of the earliest cell-to-cell signalers. As we will see later, the cells themselves produce this compound, but it can also be encountered in nature. It can cause rapid growth or hideous death depending on the concentration. At any rate, hydrogen sulfide is an intense promoter of epigenetic change. It reduces the take-up of oxygen, thereby essentially shutting down the metabolism that keeps each cell alive. Mammals overexposed to it become like reptiles. They are no longer warm-blooded. After such dosing there are a lot of things very different in their cells, including new areas of methylation.

We can suppose a reasonable hypothesis at this point: Causing violent death or escaping violent death or simply being subjected to intense violence causes significant flooding of the body with a whole pharmacological medicine chest of proteins, and in so doing changes the chemical state of virtually every cell. This produces epigenetic change(s) that can, depending on the individual, create a newly heritable state that is passed on to offspring. The epigenetic change caused by the fight-or-flight response may cause progeny to be more susceptible to causing violence. Geneticists are looking hard for a "violence" or "warrior" gene (if they are not the same), just as they are looking for a "hate" gene. At least for humans at war, there does seem to be a so-called warrior gene (the *MAOA* gene) that is found in many humans, according to a study of Finnish men who were convicted for violent crimes.[5] Its frequency in the rest of humanity is unknown still, and its presence does not predict violence. Nothing so simple as that occurs in human behavior. But there is intense study on this gene nevertheless.

The gene causes the protein MAOA to be produced.[6] This protein occurs within cells and breaks down dopamine and other chemicals. Our behaviors and our state of well-being can be derived from chemicals inside our cells. Dopamine is a neurotransmitter that protects us from depression, so having the *MAOA* gene active and suppressing dopamine might actually stimulate aggression.

In 2011, there was a well-known study of a violent Dutch family with an *MAOA* gene mutation.[7] It helped establish a link between childhood abuse and later violent behavior. Evidence suggests that the gene stays silent until exposure to violence, and then an epigenetic change turns the gene on, although one variant of the gene that makes very little of the MAOA protein.

In Finland, it turns out that the majority of violent crimes are attributable to a very small minority of the population. A large portion of the Finnish population has the *MAOA* gene (depending on race, around 40 to 50 percent).[8] But in only a few is it turned on. So, the question is: If one has the "warrior gene," why is it that in most people it is not turned on to produce MAOA chemicals?

It appears that in some people this gene is simply turned on. Perhaps, paradoxically, a first act of violence—either as recipient or perpetrator—produces epigenetic processes that switch on the gene.

The fact that the gene exists at all suggests that we either inherited it from our primate ancestors or that it has been fashioned by evolution or epigenetics. The ability to respond violently might have been necessary on the African veldt in our Pliocene and Pleistocene ancestors in order for a particular band to be able to exist among other competing bands where food sources were limited and where actual aggression for food and females was necessary.

STRESS AND THE NEXT GENERATION IN RATS

A better understanding of the links between childhood experiences and adult behavior is a major goal of the behavioral sciences. But because of our relatively long childhoods, as well as for a host of ethical reasons, it remains rare that experimental studies aimed at such understanding actually use humans as the "experimental rats." Yet some studies actually using rats are clear analogues, giving powerful insights into the profound role of maternal nurture in mammals, including humans.[9] This and similar studies underline the critical role that environmental surroundings, including the degree of parental care and positive parental emotions toward our children, have on every human early in her or his life. And not just postbirth.

In all mammals, stress molecules are deactivated by glucocorticoid receptors. The more such receptors, the more rapidly stress molecules can be disabled, thus the more rapidly the feelings we all have coming from stress—fear, worry, actual physical incapacitation, gut pains, the whole panoply of symptoms—can be alleviated. Yet the number of such important receptors is variable and can change prior to birth, depending on the environment experienced by that person's mother while pregnant. Studies[10] have shown that the fetus of a pregnant woman who has a violence-free, well-nourished, stress-free pregnancy with no exposure to alcohol or drugs experiences less *MAOA* epigenetic changes than the fetus of a woman who was subjected to high levels of violence, toxins,

and other stressors. Experiencing too much prolonged stress is analogous to having the warrior gene turned on and the stress molecules
hang around until the glucocorticoid receptors mobilize to deactivate
them.

All mammals have pretty much the same glucocorticoid receptors,
just as we all have the same kind of stress molecules, including the all-
important cortisol. Thus, in this case, the use of rats as models for people
seems to have validity. In lab rats, the clear connection between poor
parenting and epigenomic changes caused the young rats to have fewer
glucocorticoid receptors, not just in their early life but later as well.[11]
To date, this has to be a realization of prime importance to humans.

The implication for humans is unmistakable and something we all
knew to be true intuitively: An abusive childhood reduces the effective-
ness of a primary stress-response gene, leaving the abused more
vulnerable to stress, and perhaps to suicidal or murderous impulses,
later in life. But there is also evidence indicating that abuse of the young
rats, in addition to dooming them to a short and high-stress life, also
triggers the production of MAOA protein.[12]

Epigenetic study is revealing that chemical imprints laid down in
the cells of people who suffer traumatic childhoods may shape a slew of
behaviors, from depression and other mental illnesses to aggression and
perhaps even crime. It seems likely, then, that *MAOA* gene variations
have a strong effect on aggression in men who have suffered abuse. The
big question is whether that is heritable and stand-alone in its effects.
Almost all genes are "pleiotropic"—behavior is affected by more than
one gene, and genes usually code for more than one biological effect.
Scientific study has shown that fear in rodents is heritable; recall the
landmark study discussed earlier in this book of how cultivating fear in
a rat (in that case, causing a rat to come to fear a specific smell) passed
that fear on to subsequent generations. Humans have similar genes, so
how could fear, and in some cases quite specific kinds of fear, *not* be
heritable for us?

WHAT TO DO WHEN AN *MAOA* VICTIM IS IDENTIFIED IN CHILDHOOD?

Child abuse is a serious national and global problem that cuts across economic, racial, and cultural lines. Each year, more than 1.25 million children are abused or neglected in the United States, with that number expanding to at least 40 million per year worldwide.[13] In addition to harming the immediate well-being of the child, maltreatment and extreme stress during childhood can impair early brain development and metabolic and immune system function, leading to chronic health problems. As a consequence, abused children are at increased risk for a wide range of physical health conditions, including obesity, heart disease, and cancer, as well as psychiatric conditions such as depression, suicide, drug and alcohol abuse, high-risk behaviors, and violence.

They are also more susceptible to developing post-traumatic stress disorder (PTSD)—a severe and debilitating stress-related psychiatric disorder—after experiencing other types of trauma later in life. But now we know that the violence done to them creates heritable epigenetic changes that may be passed on in the form of epigenetic marks on a child's genes.

The support for this comes from multiple peer-reviewed scientific studies.[14] This research suggests that PTSD patients who were abused as children have different patterns of DNA methylation and gene expression compared to those who were not, suggesting (but certainly not proving) cause and effect. Furthermore, the researchers found that epigenetic marks associated with gene-expression changes were up to twelvefold higher in PTSD patients with a history of childhood abuse. This suggests that although all patients with PTSD may show similar symptoms, abused children who subsequently develop PTSD may experience a systematically and biologically different form of the disorder compared to those who did not suffer childhood abuse.

In human studies, child abuse has been shown to alter the epigenetic profile of the brain when examined postmortem.[15] Prenatal stress
caused by violence from the partner promotes epigenetic changes in the
DNA for this same cortisol receptor. These changes remain present in
the child's blood many years later.

THE THREE LAWS OF BEHAVIOR
AND GENETICS: ERIC TURKHEIMER

Psychologist Eric Turkheimer wrote "Three Laws of Behavior Genetics
and What They Mean."[16] These are powerful statements that are, in fact,
testable hypotheses:

> *First Law:* All human behavioral traits are heritable.
> *Second Law:* The effect of being raised in the same family is
>   smaller than the effect of genes.
> *Third Law:* A substantial portion of the variation in complex
>   human behavioral traits is not accounted for by the effects
>   of genes or families.

The upshot is that our behavior comes to us in some part by our
genome, or perhaps as important, by our epigenome. The old nature-
versus-nurture argument is no longer the correct way to posit which has
greater effect on our behavior, our genes or our environment. In fact it
is both, but in ways that were unforeseen.

Lamarckism is in the middle of this. We can have the same genes
that are found in stone-cold murderers and yet never commit the
slightest bit of violence in our lives. And vice versa. But the genes associated with major violence are like dynamite sticks: The fuse is there,
always there, and ready. What lights it is an event that occurs in our lives.
Trauma. Afterward, many individuals undergoing great trauma have a
condition that is only tardily being accepted by various militaries around

the world, one that used to be called "shell shock." During wars, soldiers no longer fit for combat were usually labeled "malingerers." But now we know that a during-life experience can indeed light the fuse and, as in all explosions, once there is a blast, there is no going back in time, no putting the fragmented Humpty Dumpty of our shattered psyches back together again. PTSD: It can cause flashbacks and memory loss (or too much memory). But the increasing understanding is that epigenetic mechanisms are involved: that the epigenome in a traumatized human is changing and adding new methylation sites or histone changes or enacting the other and probably still-to-be-discovered mechanisms of epigenome ontogeny during the life of an individual.

COMPARING VIOLENCE, PAST TO PRESENT

We now know a great deal about what happens to humans who experience or witness violence. As the history of humanity is one of war after war, it is logical to conclude that PTSD has long been a hallmark of many humans. But is our penchant for violence reducing at all over the centuries? There are arguments both for and against this proactive and hopeful idea.

A recent major examination[17] of rates of human history from the Middle Ages to now concluded that "our better angels" (presumably the parts of our consciousness that facilitate moral decisions, including whether or not to bash another human in response to a perceived affront) are, in fact, making us less prone to violence.

How many people have been in violent epochs of human history? Such a question is profoundly difficult to quantify. A series of fascinating studies looking at levels of violence within the most densely packed *cities* of London, England, through time indicates that the degree or number of interpersonal attacks per capita declined radically from the Middle Ages to the present time in London. One of the biggest surprises was that violence as measured by the *number* of person-against-person assaults was *not* confined to cities of medieval England, but took place at *higher* rates in the villages.

One way to compare relative violence from a half millennium to now is to use the number of homicides as a *per capita* figure: For every hundred thousand London inhabitants of 1500 to 1600, how many were murdered in violent, ghastly crimes? While there are many kinds of violence, homicides are more accurately reported than other assaults. New data presented by Dutch scholar Pieter Spierenburg[18] showed that the homicide rate in Amsterdam, for example, dropped from 47 per 100,000 people in the mid-fifteenth century to 1 to 1.5 per 100,000 in the early nineteenth century. Professor H. Stone of England has conducted similar studies and in them concluded that the homicide rate in medieval England was on average ten times that of twentieth-century England. A study of the university town of Oxford in the 1340s showed an extraordinarily high annual homicide rate of about 110 per 100,000 people. Studies of London in the first half of the fourteenth century determined a homicide rate of 36 to 52 per 100,000 people per year. By contrast, the 1993 homicide rate in New York City was 25.9 per 100,000.[19] The 1992 national homicide rate for the United States was 9.3 per 100,000.

When these data first came out, the high rates were blamed on the formation of overcrowded cities. Yet the surprise was that it was not the pressure of poverty and close crowding that necessarily began the descent into homicide (usually carried out by men). Most slayings in medieval England started as quarrels among farmers in the field. The knife and the quarterstaff—a heavy wooden stick commonly carried for herding animals and walking on muddy roads— were the weapons of choice. Everyone carried a knife, even women, it was noted. Given the lack of sanitation at the time, infection from even simple knife wounds could prove deadly.

Why the homicide rate in Europe began to drop in the sixteenth and seventeenth centuries is a matter of debate. The most widely accepted explanation stems from the work of Norbert Elias, a sociologist who in the late 1930s introduced the idea of a "civilizing process," in which the nobility was transformed from knights into courtiers, bringing in a new set of manners and leading to the spread of the modern state's power over the populace.[20] But another possibility exists: that there was a large-scale genetic change.

The homicide rates in Europe began to plummet in the sixteenth century and continued to do so in subsequent centuries.[21] Two historical trends were taking place over this time interval. First, because of increased transportation efficiency, people were better fed, with food getting from field to marketplace more quickly and spoilage being far less. But a second and quite different trend was also happening. It was in the sixteenth and seventeenth centuries that the recurring pulses of plague finally lessened. The wholesale death from plague that was commonplace in Middle Ages Europe finally ebbed. It certainly can be hypothesized that witnessing violent death can flood the body with stress hormones, perhaps to levels that no other undertaken or observed acts can replicate.

The act of murder in the Middle Ages would have been bloodily horrific. We who were raised in the age of television were raised on the "clean" death of a single bullet in western, gangster, and detective shows of the 1950s till now. One of the most troubling realities is that, unlike on television, violent death tends to come when a great deal of blood has flooded out of the body through a wound. In real life, most gunshot wounds are not immediately fatal, but become so from "bleeding out."

Compared to the murder weapons of the Middle Ages, our pistols and rifles are highly efficient ways to kill, and yet witnessing someone bleeding out from a single well-placed bullet in the chest cavity or through one of the major arteries of the upper arm or the upper thigh would be far less anxiety producing than watching or participating in bludgeoning someone to death, during which the attacker would end up covered with the victim's smattered brains, fat, and blood. When the assailant was not a professional soldier, and was armed with a dull knife, murder would have been intensely traumatic when witnessed for the first time.

In the Middle Ages, the weapons were crude, which made killing fairly difficult, with most violent attacks probably ending with severe wounds. On the other hand, even severely wounding someone with a club of some sort, or if in the fields—where it is suggested that many murders took place—with a scythe, sickle, crude rake, hammer, or the like, the assault itself would have been traumatic not only for the victim but for the assailant. A best guess is that most murders of this bloody and drawn-out nature were ended by multiple wounds, eviscerations, and limbs crudely hacked off from multiple attempts. As people tend to defend themselves when being murdered, especially as death in a tavern or in a field usually took place only after a face-to-face argument, the killer quite often may have suffered bloody wounds as well. This level of extreme violence causes the release of many different kinds of chemicals into one's lymph, blood, organs, and cells. Murder most foul would seem to be a prime time for epigenetic change.

Geneticists have now confirmed[22] one likely source of violence: the number of receptors in cells that are configured to receive, and in so doing defuse, the potent stress chemicals that flood the cells during a situation of intense fear or other stressful emotion (such as hate). The job of these receptors is to remove specific kinds of the free-floating but complex organic molecules that belong to the many classes of stress hormones. These chemicals are rarely produced, but when they are, they work to flood all body cells. This happens in times of great stress and fear of loss of life.

THE RISE IN TWENTIETH-CENTURY VIOLENCE

A study of crime statistics coming from the FBI also suggests that violent crime in the twentieth and twenty-first centuries in the United States has declined.[23] But that reduction is in the *relative* rate for violent crime, even though there remain periodic peaks of violence. However, the *absolute* rate has increased, because the American population has been rising steadily over the past hundred years. While the number of violent crimes per 100,000 Americans has decreased, because of the rapid increase in the American population, the actual number of violent crime events has increased. Another thing that has been rising is the way various media have become increasingly adept at broadcasting the frequency and horror of violence. Fewer people in the twentieth century were being killed over time, but ever more of us got and continue to experience the horror in progressively higher-resolution imagery. Why is this happening?

There is no doubt that the twentieth century was the period of the greatest number of human casualties in all of human history. The world population was larger, but even so, a higher percentage of the population was killed during the First and Second World Wars than ever before. What toll did these wars take on soldiers who wounded or killed their enemies? And a question never asked: What (epigenetic) toll did these wars take on the loved ones left behind?

It's interesting to look at the statistics on how many American soldiers, compared to German and Japanese soldiers, were able to pull the trigger with an enemy in their rifle sights. It is a little-known fact that a significant percentage of American soldiers were unable to kill, even at the risk of being killed themselves.[24] This was not the case for German and Japanese soldiers. From an epigenetic point of view, did millions of Germans suffer privation in the womb during their childhoods at the height of the Great Depression? Could the *MAOA* gene have impacted a great number of men and women whose normal antimurder inclinations were sabotaged by their being epigenetically programmed into killers? While crushing, the Depression in the United States was nothing compared to post–World War I Germany, where starvation, unemployment, and street violence enabled the ascent of the Nazi Party.

However, something curious happened to the offspring (the baby boomers) of the so-called Greatest Generation of World War II warriors who fought in Europe and who built war machines under time of scarce luxury, great stress, and bad food in America. In the beginning of the 1960s, the murder rate climbed radically in America and Europe, fueled largely by the children of the World War II soldiers and their sweethearts. The baby boomers became highly efficient killers, and Americans of that generation got their own bloody war in Vietnam and war in the American streets at the same time. The American inner cities burned; the Asian rice paddies were bombed and burned as well. Have our better angels brought more peace to the children of the baby boomers? Hardly.

When we look at the U.S. murder rate as calculated as the number of *actual* murders of Americans who were living in a given year, we see the true body count. The following graph was calculated by the author based on official FBI statistics.

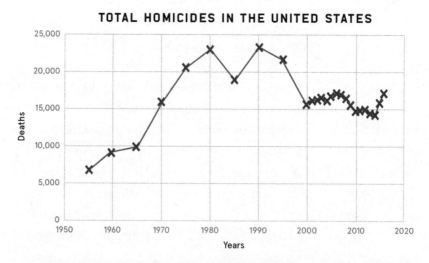

The absolute rate of homicides in the U.S. has increased in the 20th and 21st centuries, even though the relative rate has gone down. "Crime in the U.S." FBI, https://ucr.fbi.gov/crime-in-the-u.s.

In the 1980s, a generation after the '60s, crack cocaine, gangs, and many other catalysts for murder raised the body counts. Then there was a significant decrease in measurable murders in the United States in the 1990s and into the first decade of the twenty-first century. But from about 2014 onward, that trend is reversing, for the actual numbers of murders are significantly increasing again in spite of claims to the contrary. This upswing in the second decade of the 2000s is being downplayed by both law enforcement and politicians as a statistical anomaly, just as one year after another being the "hottest year on record" is ascribed by climate change deniers as a statistical anomaly. There is far more to violence than body count. A better measure might be not the cause but the effect: A probable rise in human cortisol signals more stress, which produces more crime.

2017: THE YEAR OF THE CORTISOL MOLECULE

The Chinese break astrology down to more than just our birthdays. The year in which we were born is also hugely important. There are twelve different animals representing years. The year 2017 should get its own new marker: the year of the cortisol molecule and the bloody scourge it represents. Data through 2016 illustrates the phenomenal rise in mass shootings (see page 232), the most graphic example of societal violence. The horrors of 2016, the Pulse nightclub shooting, then 2017 and the Las Vegas massacre, the horrific Texas church shooting of November, and then into the continuing horrors of early 2018. Yet, if the numbers I find are real, I predict murder and violence will continue going up and peak in 2020 before slowly dropping. We reap the epigenetically produced violence of generations past. More horror is coming. Much more, if we're to put an AR-15 in every American pot and a concealed weapon in each teacher's clothing. What could go wrong?

From 2020, if the past can predict, American murder should subside, and then rise again a generation later.

CHAPTER XI

Can Famine and Food Change
Our DNA?

E VERY discussion of epigenetics, and especially heritable epigenetics or neo-Lamarckism, invariably brings up the case of the Dutch Hunger Winter.[1] It provides perhaps the most startling, unsettling case for heritable epigenetics.

It has been thoroughly documented that a large number of German politicians and local military planned and undertook the deliberate starvation of millions of people in Holland in the fall of 1944, which preceded one of the cruelest winters of the twentieth century.

On what would come to be known as Mad Tuesday, September 5, 1944, the Dutch population celebrated, believing that liberation was at hand based on the rapid eastward advance of the Allied armies through France following the Normandy D-Day invasion of June 6, 1944. While Operation Market Garden later that September would liberate portions of southern Holland, including the cities of Nijmegen and Eindhoven, much of the northern half of Holland remained enslaved.

The Dutch government-in-exile ordered a nationwide railway strike in September 1944, but this backfired, as it only enraged the Germans, who imposed restricted access to food and medicines on more than 4.5 million occupied people. More than 20,000 Dutch citizens died of starvation. This—and the wholesale deportation of Jews from Holland, often with the support of the occupied Dutch police—resulted in Holland having the highest proportion of citizen deaths among non-Axis countries during the war.

The Germans ordered a relentless stranglehold on foodstuffs coming into Holland, as well as the export to Germany of food already in the country. The cold winter froze the canals, which normally stayed open in the winter, thus cutting off the normal barge traffic that carried much of the foodstuffs from farms to Dutch city marketplaces.

The lack of food during that terrible winter (humans need much more food in the cold) led to starvation for many in Holland and in other Nazi-occupied areas of Europe, as well as Japanese-occupied regions in Asia and the Pacific during the first half of 1945. In Holland, for example, bread, an absolutely necessary food staple, dropped from around two thousand grams per week per Dutch citizen to four hundred grams by the end of the winter and spring of 1945.

Because the Dutch Hunger Winter took place in what was a highly developed country, with first-rate medical care and scientists to track the effects on the populace not only during the famine but in subsequent generations, we know more about the role of major famine on large and especially urban populations from this event than from any previous famine. The last major famine to hit a highly urbanized country was caused by Mao Zedong in the late 1960s, and it was horrific.

At first, science concluded that mass starvation affected only those who lived through it and the children born to women who were pregnant during the time of famine. The poster child of this terrible episode was the actress Audrey Hepburn, who spent her childhood in the Netherlands during the famine and never could overcome her low weight, anemia, and chronic respiratory diseases. But most other waifs resulting from the famine had far harder lives than that of a movie star. A Dutch Hunger Winter study[2] made the surprising discovery that children from women who became pregnant after the famine was over also showed what have been interpreted[3] as effects caused by a Lamarckian change to their mothers.

The offspring of the mothers and fathers who lived through the famine showed a higher propensity for diabetes, abnormal weights from eating disorders (either anorexia or obesity), and various cardiovascular ailments that shortened the life spans of these first-generation children.

Yet it was the unexpected appearance of higher proportions of similar ailments in these children's children[4] that surprised scientists and in many ways remains one of the most singular proofs of the dark side that epigenetic change can produce. Perhaps most alarming was that the first generation of children born during or soon after the famine included an abnormally high number of children who developed schizophrenia and other mental disorders.[5]

The reason comes not from methylated DNA but from the effects starvation has on the smaller genetic bits within us, RNA molecules. Laboratory tests on starved worms showed that offspring were born with what are named "starvation-responsive small RNAs."[6] The small RNAs are a species of RNA molecule that regulate gene expression. These molecules were found to be involved in nutrition; when present, they inhibited the offspring from fully benefiting from meals. Incredibly, the starvation-responsive small RNAs were passed on through at least three generation of worms.

One of the conclusions that came from studying children affected by the Dutch Hunger Winter, rather than only lab animals, was the proclivity in print to "blame the mothers." At the time of the original research, it was the implication that only pregnant women at the time of the famine were affected, and that no negative health defects emanated from the starving fathers. After all, it was long thought that, because of Darwinian tenets, nothing affecting the health of the father could be passed on to his offspring through his sperm, and epigenetic markers accumulating during the life of the father would be genetically "erased" in the fetus. This is being overturned, with recent studies by geneticist Adelheid Soubry and colleagues at Duke University, who found[7] a correlation between fathers who were obese *before* impregnating their mates and subsequent methylation on parts of the children's DNA that codes for a hormonal growth factor necessary for normal growth. Men, too, have probably contributed to the effects of epigenetic change going forward through time.

Starvation is not the only environmental response that has been observed to be passed on in animal testing. A recent study from Scandinavia[8] showed that children living through "feast years" produced offspring more susceptible to obesity and also having shorter life spans than children whose parent—either father or mother—did not have a year when their diet was temporarily expanded to include more food or an abundance of rich food, such as meat, dairy, and cheese. This study of Swedish historical records found that men who had experienced famine in childhood were less likely to have grandsons with heart disease or diabetes than those who were well fed. Among the 1905 birth cohort were grandsons of Överkalix boys who had experienced a "feast" season when they were just hitting prepuberty, a time when sperm cells are maturing. Most of these kids died on average six years earlier than the grandsons of Överkalix boys[9] who had been exposed to a famine season during the same prepuberty window, and those grandsons often died of diabetes. When a statistical model was controlled for socioeconomic factors, the difference in life span became thirty-two years—all dependent simply on whether a boy's grandfather had experienced one single season of starvation or gluttony just before puberty.

THE GREAT CHINESE FAMINE

World War II is estimated to have killed more than 40 million humans in Europe and the Soviet Union, most in the latter. Yet even those staggering numbers do not include deaths in Japan, China, Vietnam, Burma, Singapore, and all of Southeast Asia. For those nations, there is no reliable count. However, the death toll of the Asian tragedy of World War II itself might pale compared to what was done by Mao Zedong and his various Great Leaps Forward and other draconian attempts at social engineering.

We quail at the realization of the suffering and deaths of the 20,000 Dutch in their Hunger Winter, but most neutral observers estimate that the Great Chinese Famine[10] killed 15 million to 30 million, and perhaps as many as 50 million![11] That's about 5 percent of the Chinese population that died in a three-year period at the start of the 1960s.

The initiation of the Great Chinese Famine was entirely due to Chairman Mao, who in 1958 decreed that from that point forward, no Chinese peasant farmer owned his own land. All was to be owned collectively by the state. Rigid communist doctrine demanded that not only was the land owned by the state in perpetuity but so too were the harvests (if any) from the land. The term was *collectivization*, where small farms were merged, former owners became serfs, and output was trucked to the cities without any regard to the peasants tilling the land. Even worse, the normal plowing practices were changed based on the crackpot theories of the Soviet Union pseudoscientist Trofim Lysenko, which ordered that the seeds being planted were to be planted far closer together than traditionally done, causing stunting of the seedlings, as they now had to contend and compete for nutrients.

Soviet scientists also demanded "deep plowing"[12] based on the harebrained idea that soil becomes ever more fertile the deeper one plows. Much of China is semitropical, and the soil is traditionally thin due to the rapid chemical weathering of micaceous and granitic rock minerals underneath the soil into the red clay known as laterite. Deep plowing quickly caused soil over much of the country to wash away in the torrential rains that characterize warm, near-tropic latitudes; in the large regions of China, seasonal monsoons often caused six months of unceasing rain, followed by no rain at all.

A second and concurrent decision by Mao further sealed the fate of hundreds of millions of people. Mao wanted China to be a modern country, and modern countries have large militaries. All militaries since the invention of forged iron ran on iron, and then steel, so Mao uprooted tens of millions of peasant farmers from the largely rural Chinese countryside and forcibly relocated them to cities and steel mills.

Steel production, rather than food supply, was the directive that China was required to follow.

Food, or its lack, has been a potent tool of history. Potent too in the overall well-being of humanity and even more so, as we now find, in our evolution as well.

FEAST OR FAMINE: THE MICROBIOME AND EPIGENETICS

One of the great realizations of the twenty-first century is that all animals have a diverse, numerous, and exceedingly complex series of different communities of microbes. Like all bacteria, our gut "flora" produce chemicals. These tiny chemical factories are surrounded by liquid, and they remove and put back chemicals into the liquid. It has been estimated that there can be up to four pounds of microbes in a human alimentary canal or digestive tract, with all these billions of microbes making chemicals and releasing them. Only since about 2015 has there been serious writing about what all those chemicals might be doing once they leak into cells of the body. Chemicals can cause epigenetic change, and we are now learning how much change that might entail.

Epigenetic effects on gene activity are known to occur in response to nongenetic factors such as body weight, amount of physical activity, type and amount of diet, and environmental poisons.[13] Yet the exciting if also troubling discovery is that each of those factors can affect or be affected by the gut biome. The billions of tiny microbes can turn genes on and off through epigenetic mechanisms. Many aspects of our overall health are affected by the nature of our gut biome, and this in turn can affect our mental health. Then both, of course, can affect the genes we pass on.

Microbiome is a collective term used for the genes from colonizing organisms, including fungi, viruses, and bacteria. Separate ecosystems are found in our mouth, esophagus, stomach, and along various lengths of the convoluted small intestine right down through the large. It would be analogous to a boat trip beginning in the highest parts of the Andes Mountains, above the tree line, and then moving ever eastward, through the many kinds of dry mountain forests, rain forests, savannahs, jungles, plantations, deforested angry red soil covered in new weeds, all the way to the Atlantic Ocean. The microbiome is a major player in the life of any human (or any animal, as we share this with the vertebrates, at least). Epigenetic changes can take place from sudden exposure to toxins. But what if those toxins come from one's gut? The trillions of cells in our intestines pump out chemicals that are analogous to environmental change. In this scenario,[14] our guts (or their microbes) might be the most important yet least recognized agents of neo-Lamarckian change.

The best description of this newly discovered, teeming country within all animal alimentary canals is (in my opinion) from David Montgomery and Anne Biklé in their 2015 book *The Hidden Half of Nature*.[15] They put the importance and implications to evolution and human ecology in stark terms. With an estimated 100 trillion cells living in the human body, and with each species itself the product of millions of years of co-evolution between humans and microbes as well, it should come as no surprise how important this teeming world is not only to our day-to-day lives but to human evolution as well. There is a reason our guts hurt when we are worried or stressed. But there is also a reason that our stress levels have such a huge effect on our health.

Microbes, as well as a variety of smaller organic molecules or indigenous microbiota, are essentially complex chemicals that potentially interact with the tissue cellular environment to modulate signaling pathways and regulate gene expression.

The varied microbes of our alimentary canal can be thought of as having very useful symbioses (associations with mutual benefits to each of the species involved). In this case, the microbes make possible digestion and provide a source of various nutrients: We use them as food! In most mammals, the gut microorganisms produce a number of LMW bioactive substances such as folate, butyrate, biotin, and acetate that are important in digestion. But these four chemicals can also cause epigenetic change by turning specific genes on or off.

Most pregnant women are advised to take folate supplements, but it is not just the developing fetus that needs it. This is a vitamin of profound importance to sustaining life. The efficiency of DNA replication, repair, and methylation is affected by folate availability. White blood cells, red blood cells, and other kinds of blood cells constantly have to be made anew, and this requires large amounts of folate.

Butyrate is less well known, but it has another major job to play. As we eat, we continually introduce nonbiological compounds into our body. Some of these are carcinogenic. Butyrate reduces cancer risk.

HYDROGEN SULFIDE AND ANIMAL CELLS

Seattle savant Mark Roth broke new ground in science (and possibly in cultural choices in the future) in 2008 with the astounding and paradigm-changing discovery that mice, when exposed to hydrogen sulfide (H_2S), went into a state that has been characterized as "suspended animation."[16] A more accurate description is that the mice were put into a reversible death state. They were shut down metabolically by the action of the high level of H_2S in their cells (earlier, Roth had discovered that cells make H_2S and use it as a signaler). But the higher level of H_2S allowed Roth to cool the mice to temperatures so low as to be otherwise lethal. When the H_2S was shut off, the mice came back to life. Since mice cannot talk, there is no way to know whether their brains being deprived of oxygen rendered them brain-dead.

That is the major question left hanging from the Roth results. Only later was a second question even posed: While the resurrected mice were back to doing mice things—eating, defecating, copulating, and going mad in their spinning treadmills—was there significant epigenetic change in their brain cells? The answer was that the brain was changed—and not just in ways that were considered predictable.[17] It concerns how H$_2$S relates to meat eating.

HOMOCYSTEINE BUILDUP AND HEART ATTACKS—EPIGENETIC CHANGE FROM HYDROGEN SULFIDE TO THE RESCUE

Digesting chicken breast and, above all, red meat like steak and rack of lamb leaves behind varying levels of the amino acid homocysteine. In large concentrations this amino acid has very negative effects on heart health. Its buildup is one reason that humans are not total carnivores. In excess quantities, it increases oxidative stress on many tissues. We have all been cudgeled into buying vitamins for their "antioxidant" properties, as if oxygen is a killer and maybe we should get rid of the whole thing. As in everything, however, as we metabolize and burn compounds, we use the energy released for life processes. Just as a fireplace leaves singes and smoke sludge up the flue when there is a poor draw, so too does the intense burning of metabolic processes affect cells in which the "burning" is taking place.

Two kinds of bad things happen. The endothelium, the interior layer of tissue, is degraded by too much oxidative metabolism, which is stimulated by excess homocysteine (among other compounds). Endothelial cells, belonging to the tissue, are most susceptible when they are making up blood vessels and heart muscle cells. Within cells, metabolic activities spurred by excess homocysteine from red meat cause degradation of the organelles called mitochondria, where most energy extraction takes place. These are the equivalent of diesel engines running a generator for electricity. When the fuel burns too hot, as it does with excess homocysteine, the engine itself (the mitochondria) degrades. But hydrogen sulfide is a major antioxidant and stops this homocysteine buildup.

H_2S is a product of bacteria and fungi. Only in the past ten years has its important effect on animal and plant physiology been established. Yet even less is known about its epigenetic effects. The effects of H_2S on cells are generally protective to the cell, as it causes a reduction or neutralization of reactive oxygen and nitrogen species. These effects have been reported in neurons, myoblasts, neutrophils, and macrophages. But how do animals get the stuff?

Dietary garlic has long been known for its cardiovascular benefits. Ingesting crushed garlic cloves causes a chemical chain reaction with one result being the short-term delivery of a dose of H_2S to the body, with the molecule then reacting with other chemicals in cells and causing the blood vessels to relax, reducing constriction of those vessels. This has an enormous positive influence not just on the movement of blood in and around the heart but on delivering oxygen to nerves and the brain.

We are what we eat, sooner or later. It is not surprise that when, what, how much, and why we eat what we do remains a potent source of evolutionary change in all species.

CHAPTER XII

The Heritable Legacy of
Pandemic Diseases

O N E of the most surprising details of the fabulous Napoleonic War novels by Patrick O'Brian, which began with *Master and Commander* and ended twenty-odd novels later, was how often the otherwise seemingly omniscient physician-spy Stephen Maturin "bled" his patients to reduce the anxiety of the oft-wounded Captain Jack Aubrey or as routine bleeding of the crew using an unsterilized lancet.

Bleeding was a major practice of the medical profession for centuries.[1] One would think that the lack of disinfected scalpels or lancets for bleeding would have killed off any number of the patients simply from blood poisoning and infection. Yet there appears to be a quite reasonable explanation for this practice, and this was realized well before any practice of what can be called "medicine" was invented. For a time, bloodletting was the only known treatment for the most merciless diseases of humanity: the bubonic plague and other bacterial diseases.

We see among the various plagues of humankind, from bacterial to viral, and in some cases from allergies, a potential cause of human evolution in the near past. Epigenetic changes may have been triggered not only by toxins, by ravages of disease, or by the great stress of watching loved ones die horrible deaths but also by changes in the gut biome. An epidemic is thus one of the most profound environmental changes that could spark what Lamarck considered the first step causing evolutionary change. Any plague would also produce Lamarck's second step: radical change in behavior, from simply fleeing the medieval cities to invoking spirits to seeking medical help. One treatment was bleeding the sufferer.

A great many of the most odious and deadly diseases affecting humans through history have had a bacterial origin, and the first recorded use of bleeding can be dated back to ancient Egypt; it then spread to Greece around 2,400 years ago.[2] By about 1,800 years ago, the early advocates of the medical profession, from the revered Hippocrates to Galen, extolled its virtues for literally any illness, even for obesity and unhappiness. There are references to bloodletting in writings from early Christians, Jews, and Muslims. The practice was not limited to Western medicine, as it was also used in the Americas before European contact.

It was in the Middle Ages that the practice became common for virtually anyone, and just as today a blood-test draw does not require a doctor, bloodletting was commonly done by barbers and hairdressers, with the bloodstained towel giving rise to the red-and-white-striped barber's pole. In the Middle Ages, bloodletting was the sole hope for those stricken with the most insidious and deadly of the common diseases of that time: bubonic plague. It was not until the actual scientific study of bacteria that what folklore had addressed thousands of years earlier was found to have a scientific basis: Our blood contains great quantities of oxidized iron. The element iron is something that all bacteria need. Bubonic plague is quite insidious: Its microbes feed on iron from within the white blood cells, called macrophages, of the infected person's (or rat's) immune system. Bleeding robs the bacteria of iron.[3]

When invaded by disease-causing microbes, the macrophages swing into action and carry the offending microbes into the lymph system, where in most cases the microbes are neutralized or killed. Yet the plague bacteria thrive and multiply there. This is a reason for the bubonic swelling of lymph nodes, which finally burst open as pustules. In the lymph, the plague bacteria increase in population. The only practical way to stop this was to cut off some requirement of the bacteria so that their numbers in the body could be reduced to the point that the immune system could finally get rid of them. The major nutrient needed by all bacteria is iron, so bleeding a patient of a great deal of blood reduced the iron content of the whole body. Of course, it also starved other parts of the body of oxygen. (Massive bleeding was also used by the Mesoamericans not only for disease cures but to put individuals into comalike trances—from lack of oxygen to the brain—for religious rites.)

People usually got bubonic plague through the bites of fleas that had previously fed on infected animals like mice, rats, rabbits, squirrels, chipmunks, and prairie dogs. Plague was also spread through direct contact with an infected person or animal, by eating an infected animal, or through scratches or bites from infected domestic cats. In very rare cases, bacteria from clothing that had come into contact with an infected person also spread the plague bacteria.

The many individual outbreaks of pandemic diseases received separate names.[4] The Antonine Plague (A.D. 165–80) was a pandemic of either smallpox or measles brought back to the Roman Empire by troops returning from campaigns in the Near East. It caused up to two thousand deaths a day in Rome, one-quarter of those infected. Total deaths have been estimated at 5 million. Disease killed as much as one-third of the population in some areas and decimated the Roman army. The epidemic had drastic social and political effects throughout the empire, mostly notably in Athens. The Plague of Athens (430 B.C.) appears to have been an early epidemic of bubonic plague, but because it was so much earlier than known bubonic plague events, it might have been typhus, smallpox, measles, or toxic shock syndrome (related but noncontagious). It hit the second year of the Peloponnesian War, involving Athens and Sparta. Sparta, and much of the eastern Mediterranean, was also struck by the disease.

The plague returned twice more, in 429 B.C. and in the winter of 427–26 B.C. These outbreaks included the Great Plague of Milan (1629–31), a series of bubonic plague epidemics that claimed the lives of around 300,000 people. Milan alone suffered approximately 60,000 fatalities out of a total population of 130,000. The Great Plague of Marseille (1720–22) and the Moscow Plague (1770–71) were also huge outbreaks of bubonic plague.

EPIGENETIC RESULTS OF THE GREAT PANDEMICS

Quite separate categories of epigenetic effects came from the many bouts of plague and disease that swept the human populations. In effect, plagues were a consequence of agriculture: The new bountiful sources of food led to greater populations but also to cities. Pandemics need crowded human conditions to be great killers. They need people already systemically weak, and it appears that the rise of agriculture actually reduced quite significantly the average life span as well as the height and weight of humans: Diets became more monotonous and less nutritious, and crops often failed, leading to famines. Plagues were also related to armies and war: Wars of conquest brought microbes to human populations that had been isolated and had no immune response whatsoever.

The most obvious evolutionary effect from pandemics was the lowering of human populations and the removal of large swaths of any population's gene pool. While there were the positive effects (at least from a natural selection point of view) of weeding out humans with less effective immune systems, in many smaller populations this led to severe "bottleneck effects," where the survivors disproportionately changed future, larger populations.

But a little-discussed aspect concerns the effects on survivors. It is difficult to conceive of the horror in the European and Asian cities for those who survived these pandemics. When one-third to one-half of a population (or more) dies quickly and so horribly, with unimaginable suffering, the consequence on the various human stress systems would have been significant.[5] This would have been akin to the soldiers who survive war: post-traumatic stress disorder for all concerned. Burying husbands, wives, children. The smell of the rotting dead. The loss of services. The accompanying famines as the loss of agricultural workers left fields unplowed. Broken transportation systems. And the greatest stress would have been within cities where mortality was highest.

The survivors would surely have suffered far higher rates of violence, causing and receiving, based on the levels of anger and help- lessness of so many witnessing the horror, as well as the reduction in services, including food supply. The reduction of sanitation, the rotting bodies, the increase in rats and other vermin, the incessant smell, the pollution from the burning of bodies and clothes. There would have been an increase in alcoholism. All of these would have affected levels of cortisol and adrenaline. Those changes would have triggered meth- ylation, which surely would have led to heritable behavioral changes. Survivor's guilt, but also the many ravages of mental health from seeing loved ones die, the effects of PTSD, the effects of depression. The flooding of the survivors' bodies with stress hormones on a daily or hourly basis. Stress, the increase in mutations, the epigenetic marks made, the epigenome radically changed. How could there not be enormous evolutionary repercussions for several generations after?

RELIGIOUS EXPERIENCES AND
GENE FUNCTION

The pandemics would also have opened the way for religious conver- sions. Here, too, we know that many humans undergoing a profound religious conversion or experience in life produce their own set of heri- table epigenetic behavioral changes. But another consequence, at least in medieval Europe, was a strengthening belief in many of what seemed to be their only hope for survival: help from God. Parents then surely loved their children as parents do now. The death of a child would have been no less emotionally catastrophic: Imagine seeing most or all of one's children die. The concept of heaven, that the newly dead were in a "better place," was an emotional refuge.

Intense religious experiences are common to many humans: times that seemed to take us out of our bodies, and in many instances forever changed us. Perhaps it happened just once, or perhaps commonly with prayer or meditation as we transport to another consciousness.

The kinds of consciousness change (defined by some as "religious" experiences) are diverse indeed. Until now, there was little scientific, biologically quantitative research into the intriguing possibility that specific genes are involved in whether or not a person becomes strongly religious, spiritual, susceptible to outside suggestion, or easily transformed into a deeply meditative state, among so many other episodes of what some consider transcendence. But this is increasing in scope.

It does appear that the children coming from highly religious parents become involved in the same way as their parents. Is this just cultural learning, much as the political leanings of parents, preached over the dinner table for years, usually become the political leanings of the children as well—or, as has been posited, is some aspect of this due to epigenetic change passed from a parent to offspring? It does seem that the feelings of those who feel a rapture from their faith is qualitatively and quantitatively more intense than the feelings of those at a political rally or the emotions or from emotional involvement with a sports team.

Of all the kinds of research, studies of potential genetic changes in any aspect of human brain function are among the most difficult to rigorously test. Yet there is now evidence that changes in consciousness might be associated with on-off switches controlling the production in the human cell of a protein called vesicular monoamine transporter, or more simply VMAT2. (The gene that calls for this protein to be made is also given that name, although others have dubiously named it the "God gene."[6] As always in science, other scientists denounce the possibility of such a gene at all.

There is now no doubt that religious experiences of many kinds can cause observable or measurable brain changes at various levels, from individual nerve cells to wiring to the formation of memory. They do not do so in everyone. But new work using brain scans—a method that records the relative activity of different parts of the human brain at any given time and after experimental stimulation—shows that not only can brains reveal observable changes during religious experiences but that, in different humans tested, a change of activity is observable in the same approximate region of the brain.[7]

Also as interesting is the finding that the brains of self-identified atheists show different activity patterns than those of the religious, even when there is no experimental stimulation (such as showing religious iconography to believers and nonbelievers). Neuroscientist Andrew Newberg[8] examined the brain of one self-professed atheist while that man was meditating. According to Newberg, the atheist's brain operated differently from the brains of Buddhist monks and Franciscan nuns who were also scanned while meditating. The atheist had far more activity in the prefrontal cortex, the area that produces and controls emotional feelings. According to Newberg, the atheist's brain appeared to be functioning in a highly analytical way, even when he was in a resting state. This suggests that those who self-identify as religious become less analytical during some or perhaps at all times, not just when meditating or having a religious experience.

Such religious experiences also affect memory. In many cases, those who enter various kinds of perceived religious rapture later have decreased or no memory of the specific experience. They may have a sense of time loss (the internal clock is slowed or turned off). Both reactions are associated with decreased frontal lobe activity. Some control center in the brain has been altered in a fundamental way. The best explanation is that while they retain the same DNA as before the first revelation or ultra-religious experience, a different set of gene functions is taking place, with the altered genome or epigenome increasing some proteins and decreasing others.

This is where the *VMAT2* gene comes in. This gene controls mood by regulating the production of the VMAT2 protein,[9] which then acts on the amount of mood elevators, including serotonin and dopamine, the two most powerful mood-altering drugs relating to pleasure. The conclusion among many neuroscientists from these and similar experimental findings is that spiritual tendencies involve gene expression relating to the brain's neurotransmitters.

From spirituality, then, comes a change in mood based on protein formation or the levels of these proteins in and out of cells of the body. While there is a possibility that the ability to make these proteins is purely a genetic coding process inherited and unchanging from birth, the experimental evidence says otherwise. The most parsimonious explanation is that an epigenetic change occurring in their lives causes people to have enhanced or reduced mood-altering proteins.

The new frontier is whether these are the cause of the widely believed view that strong religious identification can be genetically passed on. But is that heritability coming from unaltered genes or from genes that have markers for methylation triggers or protein formation blocks that cause the same parts of the genome to remethylate in the religious person's offspring? Spirituality serves a human purpose and over tens of thousands of years would be favored by natural selection, since strong religious belief is medically, psychologically, and socially beneficial, by extending life span and helping heal faster from diseases.

Among the most striking evidence of how spirituality, born from protein concentrations in the brain, affects not only a first generation but a second generation as well was recently documented by Franco Bonaguidi, who studied the effect of self-professed spirituality in patients recovering from liver transplants and breast cancer surgery.[10] Religious patients not only showed higher survivability from the surgeries but longer life. Yet the most striking aspect is that there is a clear heritable aspect pertaining not just to religious people begetting more religious people but to religious parents producing healthier children, based on the quantified findings that these patients were less likely to have children with meningitis.

Another heritable aspect is that patients with a strong intrinsic faith recover faster from depression than those who are not deeply religious. Depression likely causes epigenetic change in the sufferer that leads to higher levels of depression in their children. Religion can cause the opposite. Evil versus good. Stress genes versus pleasure genes on the menu for the next generation. It appears that the God gene has potential sway over at least two successive generations.

Trying to separate out the epigenetic consequences of war and violence, food and feeding, disease, and religion and religiosity as independent factors in human evolution just cannot be done. War leads to famine, disease, and quite often the imposition of a new religion. The gain or loss of spirituality. The toll of history is a mix of the many strands that individual humans have, with their lives and actions woven into a vast and complex tapestry. The designs on that tapestry are analogized by the epigenetic changes that were provoked, which in themselves led to new times and new history.

We are the product of interacting molecules producing heritable changes through many epigenetic processes. Just as we are a consequence of the vast history of life that came before us, in so many ways.

CHAPTER XIII

The Chemical Present

T HE world at present is awash in toxins that can cause epigenetic changes in humans. Some of these are the by-products of industry: pajamas that will not burn, electric stations needing chemicals to make them work, hydrogen sulfide coming up from buried garbage and organic matter, our water system contaminated by discarded pills containing all manner of powerful molecules mimicking human hormones affecting our young and old alike. But then there are the toxins we knowingly put into ourselves: nicotine, THC, alcohol, betel nut, cocaine, heroin, and on and on.

Many of us are unknowingly affected by life history events involving toxins before our time—events that happened to our grandparents. Exposures to poisonous chemicals can be one of these. Yet the degree to which our grandparents were contaminated by various epigenetically active toxins and other chemicals was orders of magnitude lower than what each of us today has encountered and will encounter. Chemicals now being absorbed from the environment and from a variety of sources are modifying DNA and controlling genes, influencing the chaos of nucleic and amino acids.

The quantitative degree to which epigenetic change is transforming our species is a study in its infancy.[1] It is clear, however, that generations living now are beginning a vast experiment in the overall evolution of *Homo sapiens* simply because the amount of chemicals present in air and water, greatly intensified in cities, is higher than even several decades ago, brought about by both advances in technology, where packaging has moved largely to plastics, as well as the great increase in human populations over these past few decades. There are a wide range of metals and organic chemicals that cause direct epigenetic change that is also associated with cancer and respiratory disease (see text box).

Of extreme concern has to be the state of global infrastructure with respect to drinking water,[2] and air pollution as global human population increases. Coal-fired energy plants are a particular danger, as they emit a variety of small particles that lodge in lungs.[3] The recent disaster of the Flint, Michigan, "water" toxicity is especially troubling. Yet those who think this is an isolated and a political rather than a biological problem are quite mistaken. In my office at the University of Washington, the water coming from the building's pipes is too toxic to drink. So too is the water in the public grade school located in North Seattle that my son attended.

This summary is an edited version from Andrea Baccarelli and Valentina Bollati, "Epigenetics and Environmental Chemicals" (endnote 1 in this chapter).

A. Metals

An association between DNA and the epigenome (in this case the addition or reduction of methylation) and environmental metals is known.

1. **Cadmium** An established carcinogen that may cause alteration of DNA methylation.
2. **Arsenic** An established carcinogen in humans but lacks carcinogenicity in animal models.
3. **Nickel** Still poorly understood but can cause hypermethylation leading to the inactivation of the expression of various genes.

4. **Chromium** A variety of genetic changes in lung cancers from chromate-exposed subjects is known, but the epigenetic effects of chromium are still poorly understood.

5. **Mercury (present as the chemical methylmercury)** An environmental contaminant and a potential neurotoxic agent that may be present at high levels in seafood. Perinatal exposure to methylmercury causes persistent changes in learning and motivational behavior in mice.

B. Organic Toxins

1–3. **Trichloroethylene (TCE), dichloroacetic acid (DCA), and trichloroacetic acid (TCA)** Environmental contaminants that are carcinogenic in the mouse liver.

4. **Air pollution** Particulate matter (PM) has been associated with increased mortality from cardiorespiratory disease, as well as with lung cancer risk. It can cause gene-specific methylation. Black carbon (BC) is also associated with decreased DNA methylation. Reduced methylation may reproduce epigenetic processes related to disease development and represent mechanisms by which particulate air pollution affects human health. Sperm DNA of mice exposed to steel plant air became hypermethylated compared to control animals and this change persisted following removal from the environmental exposure. This finding calls for further research to determine whether air pollutants produce DNA methylation changes that are transmitted transgenerationally.

(continued)

5. **Benzene** High-level exposure to benzene has been associated with increased risk of acute leukemia, which is characterized by lower rates of methylation. Even low-level benzene exposure may induce altered DNA methylation reproducing the aberrant epigenetic patterns found in malignant (cancer) cells.

6. **Hexahydro-1,3,5-trinitro-1,3,5-triazine (RDX)** RDX is a common environmental pollutant resulting from military and civil activities that has been associated with neurotoxicity, immunotoxicity, and increased risk of cancer. In epigenetic terms it changes the activity of RNAi, producing expression profiles in gene pathways related to cancer, toxicant-metabolizing enzymes, and neurotoxicity.

7. **Endocrine-Disrupting Chemicals and Reproductive Toxicants** Developing organisms are extremely sensitive to perturbation by endocrine-disrupting chemicals with hormone-like activity. Evidence from animal models indicates that exposure to these kinds of chemicals during critical periods of mammalian development may induce persistent and heritable changes of epigenetic states. Specific chemicals include diethylstilbestrol, bisphenol A (BPA), and dioxin.

THE ROLE OF TOXINS

The cause of Parkinson's disease, one of the great scourges of humanity, has been attributed to environmental toxins.[4] The nervous and muscular systems of patients are usually compromised, and early in the progression of the disease patients usually try to hide their shaking hands and involuntary muscular problems. It is an inherited disease, and more than a dozen genes are implicated in the actions (and inactions, depending on the gene) that are involved. A great deal of research has been done and enormous quantities of research money spent. So far, less than 10 percent of cases can be tracked down to actions of the dozen genes involved, and 90 percent are attributed to some aspect of the environment instead of genetics.

Environmental toxins (see text box on page 201) range from a seemingly simple thing such as too much manganese, a metal used in the steel industry, to common pesticides entering a person's body in too great a concentration. Too much lead affects nerves, mental processes, and memory, and causes degeneration of the nervous system.

While most researchers who want to understand human diseases go the white rat or mouse route, another intensively studied species is a small worm with the long name of *Caenorhabditis elegans*. In one study, this tiny invertebrate was subjected to high concentrations of manganese[5] in order to get evidence about the effects of this environmental toxin on hormonal response. The fact that an invertebrate so evolutionarily distant from humans shares many chemical triggers and signalers shows how ancient and conserved many genes are, especially those involved in response to environmental triggers such as hunger, predation, poison, or temperature extremes. The human hormone dopamine, present in all mammals, is even more ancient, and it is present in *C. elegans* as well. In response to poisonous levels of manganese (a metal that all human workers in steel plants and other industrial concerns are exposed to), the tiny worm's body was flooded with dopamine.

Researchers noted that the steps were: (1) too much manganese; (2) body flooded with dopamine; and (3) effects mimicking Parkinson's, as well as reducing life span, and causing oxidative stress in other physiological functions. The end result was degradation of the nerves.

MICHAEL SKINNER AND
AGRICULTURAL TOXINS

Modern civilization produces a diversity of chemicals entering the biosphere through many avenues. Watchdog government and medical agencies have warned us for decades about the more egregious pollutants that can affect human health, such as dissolved lead, arsenic, manganese, and other metals, as well as airborne pollutants; what is new is the increasing danger from exposure to many kinds of pollutants that prior to industrialized civilization were either at low concentration or did not exist at all in a natural state. These can certainly cause epigenetic changes that are heritable in laboratory animals. Many of these are not toxic elements included in products but are organic (carbon-bearing) molecules that can be found in modern industry, such as fire retardants in children's clothing and chemicals such as PCBs, which are used in energy transmission, among other things. Regulatory laws have tried to protect water used in human consumption and water used in agriculture, as well as the planet's water reservoirs in lakes, rivers, oceans, and ice caps. Yet chemicals find their way into each, and many come from agriculture.

North America appears to be one of the most polluted regions on Earth, in spite of it hosting three of the world's most productive countries in terms of manufacturing and agricultural output—or is this the cause? The chemicals that can now be found in both aquatic and land animals, including humans, and in human mothers' milk attest to this.[6] So do a number of alarming new studies, none more so than the 2017 study of North American males' sperm counts.[7] Based on tens of thousands of samples from men in North America, South America, and Africa, the study showed a huge drop in the sperm numbers and normalcy in North American men over the past four decades. The same was not found in men from South America and Africa. There is no proof that this is caused by toxic chemicals, but it seems a likely possibility.

Our large human population obviously needs a lot of food. To reap adequate harvests, the large-scale industrialization of agriculture has increasingly relied on artificial fertilizers, as well as herbicides for killing weeds, fungicides for dealing with fungal growth and the destruction of plants, insecticides for crop-eating pests, and hormones to grow food meat faster.[8] It appears that far more scientists are engaged in the discovery and production of such chemicals than there are warning about them. Michael Skinner of the Washington State University and his numerous colleagues have used multiple studies[9] to show that when rats are exposed to even low levels of commonly used agricultural chemicals, there are multiple generations of change. This is where heritable epigenetics and the recent finding of the drop in the sperm count of North American men come into play.

In his studies, the Skinner group exposed rats to several of the most common of agricultural chemicals. One of these is a fungicide called vinclozolin. Among the laboratory animals (rats, in this case) exposed to this chemical in the Skinner studies were pregnant females. In the subsequent generation born from the exposed females, the males, upon reaching sexual maturity, showed sperm counts that were lower than from males born of unexposed females. The rat sperm that was produced in this first generation was also often morphologically deformed. Significant as this finding was, the greater discovery was that the next *two* generations of male rats derived ultimately from the exposed females showed the same male reproductive damage. But there was no DNA change in any of them—only a heritable epigenetic change moving through time. Their genome was not affected, but their epigenome was. The exposure of the first generation of rats produced methylated sites on their DNA that were new, and passed on.

"But rats are not humans!" critics of these studies wrote. "And besides, one can bypass all these problems by eating organic!" Hardly. Modern humans are now the rats, and we can assume that we are passing on many such epigenetic traits to humans not yet born, coming from the products used on the food we eat and from so many other chemicals of industrial might.

From there, Skinner and his group tested ever more kinds of industrial chemicals, some major bestsellers for the chemical industry, including chemicals sold to the public that led to diseases in the reproductive and immune systems as well as the kidneys. These diseases would show up in successive generations, according to Skinner's work.

REEFER MADNESS

Some of the chemicals that appear to be triggers for epigenetic change are those that humans willingly (or, in addiction, unwillingly) partake in: tobacco, drugs, and perhaps alcohol.

Cannabis (with its main ingredient THC) is being legalized in an increasing number of countries, including parts of the United States. In both Washington State and Colorado, marijuana has been legal for some years, and pot shops are everywhere. Driving while impaired by pot and alcohol is now blamed for the rise in fatal car crashes in these states. But the most significant aspect is that pot is now far easier to obtain and has effectively been given a seal of approval in the minds of the young. So ever younger users are involved in a large-scale medical as well as social experiment.

Marijuana is the most commonly used illicit drug in the United States, and it seems to be getting more popular by the day as it is becoming legal in an increasing number of states. According to a 2012 survey done by the U.S. Department of Health and Human Services, an estimated 18.9 million people have used it sometime in the past month.[10] That's up from 14.5 million in 2007. As more people use it, more people seem to think it's safe. That same survey showed that, in 2007, 55 percent of kids between 12 and 17 years of age perceived "great risk" in smoking pot. In 2012, only 44 percent did.

A few years ago, one study[11] showed that adolescent rats exposed to THC (the molecule primarily responsible for pot's mind-altering effects) are more likely to self-administer heroin as adults than are rats not exposed to THC. This pointed to weed as a "gateway" into other kinds of addictive drugs.

A subsequent study[12] aimed to see whether any of these effects carried into the next generation. Over the past decade or so, many researchers, including Michael Skinner's group, have reported that a wide variety of environmental exposures leave chemical marks on DNA that get passed on for several generations. To see if marijuana exposure might do the same, male and female rats were periodically injected with THC throughout their adolescent period. This pattern of exposure was meant to mimic the typical pot-smoking teen.

Several weeks after the exposure ended (enough time for all traces of THC to disappear), the researchers allowed the animals to mate. Immediately after delivery of their pups, the offspring were transferred to another cage to be raised by a female rat that had never been exposed to THC. When those babies reached adulthood, even though they themselves had never been exposed to THC, their brains showed a range of molecular abnormalities. They had unusually low expression of the receptors for glutamate and dopamine, two important chemical messengers, in the striatum, a brain region involved in compulsive behaviors and the reward system. What's more, brain cells in this region had abnormal firing patterns, the study found.

This second generation had altered behaviors as well. Compared to the control groups, rats whose parents had been exposed to THC were more sensitive to novelty in their environment and were more likely to self-administer heroin by repeatedly pressing a lever.

OTHER DRUGS

Environmental factors capable of inducing epigenetic modifications include exposure to certain chemicals, but there are many others that governments allow, such as one of the world's greatest killers, nicotine, which, among other things, affects fertility through epigenetic pathways.[13]

As do "hard" drugs, from cocaine and opioids,[14] to ones even more prevalent in Asia, especially the nut of the *Areca catechu* palm tree. So-called betel nuts are used by a billion people. The children of the many who chew betel nuts experience higher-than-average incidence of depression, attention deficit disorder (ADD), and lower-than-average intelligence.[15]

The association between increased risk of disease and inherited epigenetic modifications is also particularly conspicuous in the case of betel nut usage. The latter produces rapid morphological change in its users: Activation of the drug, which is derived from a small tropical nut growing profusely through the Asian and tropical Pacific region, requires that it be mixed with small bits of coral lime. The result is a red liquid that is eventually expectorated (creating what looks like spilled blood on the roads and sidewalks of the Asian and Pacific megalopolises). The user's teeth are ground down to sharpened spikes as well as being permanently stained blood red. This creates a second suspected epigenetic change. Without molars to grind grains, much of the food being consumed by advanced-stage betel nut addicts is not being properly digested. The chronic malnutrition of usually lower-income groups (the largest segment using the drug) causes inflammation of the digestive system, thereby affecting intestinal flora.

As with any drug use, it is the user who is most affected—at least according to current social theory. The undeniable and highly negative effects on social relationships of an addict, especially with family, create their own devastating mental and physical problems. Now it is becoming clear that epigenetic effects are passed to children of both men *and* women users. The results in some children with drug-addicted parents: a higher incidence of high blood pressure, diabetes, and changes in metabolism affecting digestion, formally named "metabolic syndrome."

Betel nut studies are among the first studies to show the contribution of males to problems in their children.[16] This is now being demonstrated for fetal alcohol syndrome as well.[17] Men have long been given a free pass in a "blame the pregnant mothers" sentiment of centuries' usage, perhaps the ultimate misogyny perpetuated by generations of (male-only) doctors and scientists looking at fetal problems.

THE POSSIBILITY OF A NEW PURITAN AGE

In 2014, a provocative and rage-demanding photo was used to tout a new bar. The photo showed a late-term pregnant woman smoking and drinking. She also looked as if she was high on lots of substances. The caption read: "Gestations!—New York's first bar for pregnant women."

This advertisement for a bar catering to pregnant women turned out to be a hoax of a very, very bad idea.[18] But could a realization that it is not just pregnant women who put developing children in the womb at risk? What might be the reach of heritable epigenetics in legal and ethical restrictions on an individual's freedom of choice?

London in the earliest part of the seventeenth century was a hotbed of plague. But between the Black Death outbreaks was the golden age of early Elizabethan theater, including the heyday of Shakespeare and his Globe Theatre. But only two short decades after this exuberance of life among Londoners, the theaters were all shut. The cause: the ascendance of a group of religious fundamentalists known as the Puritans, who eventually became powerful enough to bring about the execution of King Charles I, as well as dragging all of England into its own revolution under Oliver Cromwell. In a very short period of time, the mores of a large segment of an entire society shifted, and the Puritan ideal allowed no drinking, dancing, and on and on.

How much will society in the near future value not only the unborn, developing child but the children still to be conceived? And can a wider understanding of epigenetics, disseminated in a new way and accepted as a reality of evolutionary change, including to human behavior, produce unintended consequences?

CHAPTER XIV

Future Biotic Evolution in the CRISPR-Cas9 World

U N T I L recently, tracking human evolution in the distant to near past was the province of paleoanthropologists—for it was only through the study of ancient bone morphology that science could track evolutionary change. But the revolution in studying evolution unleashed by the many potent DNA techniques for studying the present and past genomes has opened a whole new world of understanding about the "how," "where," and "what" of recent human evolution. The surprising bottom line is that not only have we been doing some major reshuffling of the human genome since our species formation, but it appears that the rates of human evolution have been increasing over the past thirty millennia.

By 25,000 years ago, humans had successfully colonized each of the continents, save Antarctica. Many islands were still "waiting" for humans. Adaptations to the many locales led to what we now call the various races of humans. While it was long thought that such obvious features as skin color were purely adaptations to varying amounts of sun, more recent work suggests that much of what we call "racial" characters might simply be adaptations brought about by sexual selection, rather than adaptions for survival in various environments. But many other adaptations, most invisible to morphologists, were happening as well.

With the globe fairly well populated (with a few late invasions of such larger islands as Madagascar, New Zealand, Polynesia, and Hawaii), one might expect that the time for evolving would be pretty much finished. But that turns out not to be the case.

NEW STUDIES SHOWING RECENT EVOLUTION

What is the rate of recent human evolution? A study by Henry Harpending and John Hawks has given a dramatic answer, updated in 2010 in a book by Harpending and Gregory Cochran. (These authors have also attracted a fair share of controversy as well.)[1] Their more scientific findings suggest that over the past 5,000 years humans have evolved as much as one hundred times more quickly than any time since the split of the earliest hominid from the ancestors of modern chimpanzees some 6 million years ago. And rather than seeing a reduction of evolution of those characters that, combined, are used to distinguish human races, until very recently the human races in various parts of the world have become more distinct, not less. Only in the past century, through the revolution in human travel and the more open behavioral attitudes of most humans to those of other races, has this trend seemed to have slowed.

To arrive at this conclusion, researchers have analyzed data from the international haplotype map of the human genome and genetic markers in 270 people from four groups: Han Chinese, Japanese, West Africa's Yoruba, and northern Europeans.[2] They found that at least 7 percent of human genes have undergone recent evolution. Some of the changes were tracked back to just 5,000 years ago.

One would think that genetics would have much to say about the origin of human "races." But there is a strong dose of political correctness providing headwind for any scientific project that uses the word *race* at all, and a popular 2016 account in *Scientific American*[3] suggests that the word is scientifically meaningless, which echoes prior suggestions in peer-reviewed journals.[4]

Nevertheless. There are certainly examples of the evolutionary changes affecting people in different parts of the globe, thus between human populations that have little genetic exchange. It has been noted that in China and most of Africa, fewer people can digest fresh milk into adulthood[5] than in Europe or North America. Yet in Sweden and

Denmark, the gene that makes the milk-digesting enzyme lactase remains active, so almost everyone can drink fresh milk. This may explain why dairy farming is more common in Europe than in the Mediterranean and Africa. Other cases include the evolution of lighter skin and blue eyes in northern Europe and partial resistance to diseases, such as malaria, among some African populations.

Other studies have discovered evidence for recent change due to natural selection rather than random mutation—in other words, evolutionary change to improve the fitness of various geographic populations of humans. The kinds of evolutionary change discovered by one study[6] included resistance to one of Africa's great scourges of humanity, the virus causing Lassa fever.

Yet while these studies seem to reaffirm that we are not yet finished being first-class evolvers, others take quite a different tack. It is clear that modern medicine is very successful at keeping alive individuals who would otherwise die before reaching sexual maturity. The large numbers of premature babies are but one example, and while such early births may be unrelated to genetics, they are certainly evidence that technological humanity is impacting survivorship—itself the driver of evolution. Some evolutionists point to this and the near absence of human predators (another common driver of evolutionary change in natural prey species) among many other aspects of natural selection no longer applying to humanity. But if not natural selection, inevitably there will be "directed human evolution." The most important new tool allowing us to drive our own evolution is the already described and revolutionary gene-insertion technique known as CRISPR-Cas9. It is the very novelty of the technique that renders obsolete so many prior speculations about future evolution.

CRISPR-Cas9, discussed earlier, is the odd name given to a group of DNA sequences (and now a technique used by humans) that first evolved in bacteria, and probably did so long ago, soon after bacteria first appeared. These are lengths of DNA that contain smaller pieces that originated in viruses but were then inserted into the bacterium during a viral attack. But the bits of DNA are used by the bacterium in a way analogous to an enemy capturing the weapons of its foe and then turning those very weapons against their makers. In bacteria, the viral DNA becomes a way to search out and destroy similar DNA if the bacterium is attacked by the same viruses. Warfare seems to be another definition of life—warfare of attack, kill, and turn the victim into a virus-making machine. But if the bacterium survives, the attack weapons become a prime part of the bacterial defense system.

It is this bacterial defense system that has been brilliantly copied by geneticists to produce the most important new biological weapon against a variety of genetic diseases. CRISPR-Cas9 is genome-editing technology that can make permanent modifications of genes within targeted organisms, be they humans with disease or food material that is being "improved." Any such engineering is an environmental event that is by definition Lamarckian. More powerful follow-ups to CRISPR are now being constructed, and no longer by academics but by biotech companies.

The use of genome-editing technology is increasing in biological labs around the world. There are profound problems, not least of which is that a gene placed into an organism has to have every cell make the change. Thus, in an organism that already has divided into multiple cells, the CRISPR process will hunt down specific genes and snip them out. But there is no guarantee that the targeted gene will be edited out *everywhere*. This is the major problem of trying edit genes in a living, post-embryonic human. The treatment might still work if a sufficient number of cells have the change made, but perhaps not.

CRISPR-Cas9 does have great promise and upside. Potential uses include:

1. Helping remove malaria as the deadly threat that it is to humans in the tropics and subtropics by creating mosquitos that are immune to the parasite that causes the disease

2. Helping cure a patient's cancer by altering immune system T cells
3. Treating muscular dystrophy
4. Building bigger animals (more meat) or new kinds of plant crops
5. Engineering organs grown in pigs to be used in humans
6. Helping defeat diseases such as HIV

But for all the good, there is always the dark side.[7] It is logical and probably necessary in the dangerous world of aggressive nuclear-armed nation-states of the twenty-first century to make every effort for "defense," when in reality most weapons are about "offense." Biological weapons in the form of human disease germs have been around for almost a century. But never before have humans been able to design better weapons based on animal morphologies and then, one gene at a time, transform them into something more lethal.

The gene-altering technique of CRISPR when applied to microbial-sized bioweapons has certainly already been attempted or perfected.[8] The most virulent viruses are Ebola Zaire (a variety that kills the majority of those infected); Marburg virus; rabies; HIV; smallpox; hantavirus; certain strains of influenza (such as any new variant as powerful as that of the 1918 pandemic that is estimated to have infected 40 percent of humanity and killed more than 50 million of us); dengue, a miserable tropical disease called "breakbone fever"; and rotavirus, which causes alimentary tract diseases.

Imagine if these most horrifying of human diseases had the infection virulence of the common cold, whose virus can stay alive and ready to infect even after several days on a tabletop or door handle. Or if they could be passed through respiration of infectious particles that have been coughed or sneezed into the air of crowded human populations. These would be rapid attack weapons.

But more insidious would be mating high virulence to diseases that require enormous societal outlays for care, where significant proportions of a nation could be infected with long-term, devastating diseases such as HIV varieties as yet nonresponsive to any known treatment. In analogous fashion, bioweapons that have been engineered into existence could target various plant crops or food animals of an attacked nation. These could be potentially untraceable, which is why various countries are trying to set up a database that would (hopefully) be able to identify the source of the bioweapon in a manner analogous to how various kinds of nuclear material can be traced back to their source breeder reactors.

THE DOGS OF WAR

Conflicts involving armed humans fighting each other remain the most common type of human warfare. And thus the ongoing effort to enhance humans into "supersoldiers." None exist, but methods for creating them are not at all science fiction and theoretical. China's production of not one but two genetically modified "Frankenbeagles" has brought home to many in and out of academia that we are already in a new reality.[9] Chinese scientists have taken one of the friendliest, least menacing dog breeds and doubled its muscle mass. By making two, the Chinese were telling the world that this was not a freak chance but an intended result. The threat is quite clear: Take a rottweiler or German shepherd or pit bull . . .

At Guangzhou General Pharmaceutical Research Institute, these two beagles were grown from embryos that had each undergone editing of genes used to dictate the amount of muscle in the dogs. Full-sized, they became massive, bulging dogs with happy faces and menacing builds, ideal for police and military use.[10]

The Chinese have made it a scientific priority to become proficient in gene editing to produce "designer animals." The list so far also includes goats, rabbits, pigs, rats, and monkeys. It is the latter that will produce the needed experience to jump to humans. It is taxing work in that at the moment (2017–2018), with current CRISPR methods, a large number of embryos have to be used to yield any positive results. For example, the Chinese "edited" (used the CRISPR-Cas9 method on) sixty-five embryos, from which twenty-seven dogs were born, of which only two ended up with the doubled muscle mass. But this research is just beginning.

These results are not just "one-offs" in that there are now breeding pairs of various kinds of modified dogs. Because the editing occurred in embryos, the genes affected (it was the mutation of a single gene, called myostatin, that caused the change) change the germ line. So any puppies that these dogs produce would have this gene changed as well. Very quickly, if not already, a new race of dogs would be produced.

One single gene, which is all that CRISPR-Cas9 is currently able to affect. But in many cases, there are single genes that control other genes. One gene editing is fully capable of creating many kinds of "Franken-animals" that have many other altered traits than just muscle size. The animals in these cases might be candidates for genes affecting intelligence, for instance. And even though this editing can be done only one gene at a time, there is no reason that many, many individual genes cannot be edited within a single embryo at a time.

One question that was not asked by the many reporters writing about these first gene-edited dogs concerns the source of the funding. Was any of it from military funds? Man's best friend has a long history of fighting human wars alongside human soldiers. The dogs are also soldiers themselves. And it is not just in the past. The American raid that killed Osama bin Laden and others of his family and entourage had a helicopter-transported canine soldier whose name and even breed have been shrouded in secrecy ever since. Some have been (and are) trained to kill humans or other dogs. Some have been used to find explosives, serve as keen sentries, as scouts, as communications links, and certainly as mental health service dogs for morale.

Or the opposite of that: Dogs can cause intense fear—fear that can be greater than even the terror of facing well-armed soldiers. Now there can be literally new breeds of dogs that are smarter, bigger, more savage, and less prone to fatigue—with a superb sense of canine smell.

THE JUMP TO HUMANS

If muscle mass can be doubled in dogs, the same could be done to humans. This is not lost on the U.S. military or its scientific weapons division, the Defense Advanced Research Projects Agency (DARPA). According to a 2015 edition of *Tech Times*,[11] in 2013, DARPA launched a solicitation effort for a project known as Advanced Tools for Mammalian Genome Engineering. More than just being focused on the engineering of any mammal's genome, the project was specifically launched for the bioengineering of humans. The official proposal page for this well-funded project was explicit. It sought competitive proposals from researchers for military uses that increase the efficiency of bioengineering bioweapons and bioweapon defenses.

So what can be done in humans? The rapid introduction of large DNA segments with many genes into human cell lines would allow bioengineering of the humans species. Certainly the characters and genes allowing more efficient and deadly soldiers is being looked at the world's militaries.

The limitation of CRISPR at the present time is that it is a one-gene-at-a-time system. The American military wants something far more powerful: the next generation after CRISPR. They want to produce things that biology never produced, and they intend to do it by this most Lamarckian of mechanisms: gene insertion during the life of the organism.

HOW CLOSE ARE SUPERSOLDIERS?

It is hard to get past the smoke that surrounds such a project. The first and most glaring question comes from the Chinese work on dogs. Does the U.S. military really think it could do the equivalent with seventy or so human embryos, bring as many of them as possible to term after implanting them in willing women, and, of the half that survive, have two—a male and a female—that make it to supersoldier status? How and where could this conceivably be done? On the other hand, what is to prevent us from embarking on a road that could lead to a new human species, one swapped-in gene at a time?

Yet those profoundly practical and moral questions are ignored in the hype. This, from something called the *Activist Post*, points to what traits a supersoldier might exhibit in terms of military upgrades:

> Smarter, sharper, more focused and more physically stronger than their enemy counterparts these soldiers will be capable of telepathy, run faster than Olympic champions, lift record-breaking weights through the development of exoskeletons, re-grow limbs lost in combat, possess a super-strong immune system, go for days and days without food or sleep . . . Then there's the emotional side. These soldiers will have the empathy genes deleted and show no mercy, while devoid of fear . . . Even more disturbingly, the "Human Assisted Neural Devices program" involving brain control allows the "joystick" remote operation of soldiers from some far away control center.[12]

Research into this topic is repeatedly mentioned when there are discussions about how to improve combat soldiers while also reducing their risk of mortality in combat. The ideal case would be that no human soldiers at all are used on the actual front lines. All could be done from behind the lines, using drones overhead and without pushing soldiers into battle. From self-driving cars to self- driving and fighting tanks, ships, planes—it is all of a piece.

There have been many magazine-selling articles about what a "supersoldier" would be like, from enhancements of biological functions (the need for food, water, rest) to actual physical changes (larger, more muscles, etc.) that alter human anatomy so radically as to produce organic armor or "telepathy" and its ilk. Even speculation about producing humans that can last longer on less food and water would involve enormous biological and genetic changes. The commonalities between the many "news" articles about this sensational and emotionally charged story are that almost all concentrate on physical changes (like the muscular dogs) when the surely most effective changes that could be engineered into any "supersoldier" would be behavioral. These could be realized through gene alterations that would have a huge upside for any military sending its soldiers into a war zone. The ultimate killing machines: humans without empathy or fear or normal reaction to stress molecules. These would be the changes that would win wars; these would be the gene alterations that would not be seen by the public in the making. No bulky Arnold Schwarzenegger. Rather, normal-looking humans who lack the emotions that stop us evolutionarily produced humans from being killers.

Many countries with advanced militaries are taking note of the process. For instance, in autumn 2017, Vladimir Putin of Russia said that the impending reality of genetically produced supersoldiers could be "worse than a nuclear bomb."[13]

The fear-driven sensationalism of the whole supersoldier debate is actually a metaphor. There are far worse problems that can arise from gene-editing than single soldiers, no matter how "super" they are. Yet that this conversation dominates the subject simply reinforces the unease that not only individuals, not only scientists, but also entire governments have about a technology that could easily turn on and bite its inventors.

THE NEW LAMARCKIAN AGE
OF GENE CHANGING

The advent of the CRISPR–Cas9 technology becomes a kind of throw-back to the period from the mid-1940s into the mid-1950s, when American and British scientists lost the battle for controlling the atomic bomb to politicians. The scientists were so naïve: They believed their calls for peace and disarmament would sway the Allied powers. They believed they could keep the nuclear genie in the bottle of scientific control and scientific committees while scientists on both sides of the Iron Curtain rushed to make the atomic bomb much more destructive by using this first design as the igniter of thermonuclear bombs, the fusion of hydrogen bombs.

Nuclear fission was presented to the world as a tool. And indeed the nuclear power plants that have successfully supplied a nontrivial percentage of global energy show what a powerful tool it was. CRISPR is also presented as a tool. But it was academics that brought about nuclear fission. Only long after did industry figure a way to make money off splitting atoms. Now that equation is reversed. All American scientists know that, since 2010, research grants from the traditional funders, mainly the National Institutes of Health and the National Science Foundation, are ever harder to get. But biomedical companies are flush with cash from the prices of drugs already at market. So too with gene editing: Corporations have quickly overshadowed research institutions in its use.

The whole conversation can be summarized as follows: the process sometimes referred to as "gene editing" could end up being one of the most significant health technology ever introduced, and a diverse suite of global companies are racing to use it to combat genetic diseases (inherited or acquired). CRISPR makes three categories of DNA alterations possible: embryonic modification to eliminate genetic disease, alterations to protect against future disease, and genetic enhancement of human form and function.

In this technological present, the pharmaceutical industry is doubling down on CRISPR for novel drug development. At present, any therapy based on CRISPR technology would have to involve three steps: remove cells from your body, alter the DNA, and then reintroduce the cells into your body.

CRISPR holds the promise of transforming the human species. But of greatest ethical concern is so-called germ line engineering, permanent alteration of DNA in sperm, eggs, or embryos, which from that point going forward have a different set of genes, a set that will be passed on to progeny.

The CRISPR-Cas9 method is a tool, one that is being used in agriculture, for medicines and new drugs, and hopefully for curing some of the most horrible scourges of humanity, genetic diseases. It is a tool that works using an epigenetic process, and thus a neo-Lamarckian tool coming from defense mechanisms of prokaryotic cells. But it is also an invention that has now been around for more than a decade, but that also needed to be tested successfully on what many scientists and physicians consider its most important subject: humans. Thus it was an important milestone in 2017 when American scientists became the first to successfully gene edit a live human embryo.[14]

THE GENIE IN THE BOTTLE

Still unknown is whether inherent limitations of CRISPR will keep it from being a biological danger—and whether scientists will successfully partner with governments to make wise decisions on the troubling ethics and equally troubling potential dangers of the method. *But it is just the vanguard of gene-editing tools.* Soon the one-gene-edit limitation of CRISPR-Cas9 will seem as primitive as the first airplane is to a modern stealth fighter. There is too much money pouring into this field for this not to happen.

More power for good can come when it becomes routine to be able to discover, target, and then change patient-specific genes that in a person may be contributing either to a specific disease or in the inability of that person's immune system to fight the disease because of specific drug resistance.

The for-profit biomedical industries want to move the current technology many steps forward. Yet, as most biologists familiar with its uses strongly advocate for some kind of international control, there are already cases circumventing what little control already exists.

And how about regulation? In 2016, a gene-edited CRISPR mushroom[14] escaped U.S. regulation and now can be cultivated and sold without further oversight. Not long after, the U.S. government allowed a new type of corn genetically modified with CRISPR-Cas9 to go into production; the degree of scrutiny by the government in this case was minimal, and with the 2016 presidential election of a man intent on elimination of "regulations" and totally oblivious to any aspect of science, there are good reasons to wonder how far regulations will go—or, in this case, not go. In the 1940s, the inventors of the atomic bomb tried to keep control of their toy. They failed.

So, what might the dangers be? It is that all-too-real genes, like imaginary genies, do not stay cells at all times. Sometimes they "jump." A real danger is if some newly created heritable gene mutates once inserted into an organism, be it manifest in a plant, animal, or microbe. There may be a greater possibility of what some specialists have dubbed a "gene-editing catastrophe" coming from attempts to modify plants than one coming from the dangers posed by modifying human genes, simply because of our global reliance on plants for food, and the lesser scrutiny on plant modification (as well as the secrecy of agricultural companies hoping to profit from some new variety of corn, wheat, or other food crop.)

In the past, the world's universities produced but a handful of scientists capable of working on nuclear weapons each year. The single greatest difference with that dangerous tool (nuclear fission) and the new tools of gene splicing is that there are literally millions of biologists produced by the sum total of global universities capable of using CRISPR. A book was recently published showing how "you at home" can run your own experiments using CRISPR technology. CRISPR has the potential to benefit humanity, but the neo-Lamarckian gene-editing programs also could kill a large number of people, either as a weapon or as a well-meant food source or disease cure run amok. No one opted for Three Mile Island or Chernobyl—other human constructs gone rogue.

Looking Forward

How much of our own decision making, speeches, and declarations of independence have been influenced by our upbringing, and how much is based on levels of the various powerful and changing hormones within us that have been brought into our own genomes by heritable epigenetics?

In this book, the case has been made that more than now accepted we are influenced by the interplay of genes brought into life by natural selection, but perhaps more specifically by genes altered by epigenetic processes. These include methylated genes and altered histones, along with the effects of small RNAs that became heritable through some neo-Lamarckian event in the lives of our parents, grandparents, or great-grandparents. Chapter 10 tackled the notion that humanity over the past few centuries, and more specifically in the past few decades, has sociologically evolved, strung along by our "better angels." More accurate seems to be that we have evolved by epigenetic devils.

Virtually every human on Earth lives in an environment where we inhale, ingest, or come in physical contact with a greater volume and greater diversity of biologically active environmental chemicals than in any previous time period. As noted earlier, a 2017 study of men living in North America demonstrated that their sperm counts were significantly reduced over the past forty years.[1] The authors of this study proposed only two possibilities: that the chemicals so prevalent in air and water in North America have caused a reduction in reproductive ability or that it is the rise in average summer temperatures in North America that have caused or contributed to this. It is known that in all mammals sperm need a lower temperature and are thus stored in testicles rather than in the confines of the bulk of the mammalian body. For the past decade, a quixotic scientist in my home state has tried to warn of what pesticides and herbicides are doing to us: not as Chicken Little but as an explorer of what these chemicals might do to future human evolution. Michael Skinner looks for the processes that the chemicals he studies might reveal, as well as the dangers that those same chemicals pose. For that, it appears from afar that he has been a target of criticism from the agricultural industry.[2]

We live in a world with more humans than at any time in history, and every human contributes to carbon dioxide in the atmosphere and to some contribution of chemicals that can affect biology. We live in a climate that is changing as fast as at any point in deep time.

We are all the products of evolution by Darwinian mechanisms. But we may also be, to some extent, the products of the environmental stresses that happened on given days in the lives of our parents and grandparents. This occurs through heritable epigenetic processes triggered or initiated by environmental change.

What could possibly create such environmental change in the future? Here is a list, each with physical changes and societal effects that might happen as a consequence.

1. *Stress from loss of habitat and agricultural yield from current farms located at sea level or below.* Sea level rise at even the most conservative rates projected by the latest Intergovernmental Panel on Climate Change[3] poses the special dual threat of soil inundation coupled with horizontal salt intrusion into the world's major river deltas, such as the Nile, Mississippi, Mekong, Ganges, Fraser, and others. These low-lying sedimentary assemblages are susceptible to damage by storm-wave erosion (through physical removal of sediment, as well as the biological destruction of rooted green plants and trees that are the primary physical framework holding together the delta sedimentological package).

The acreage of productive agriculture at the present time is vast. The agricultural production from farms at or below sea level currently feeds a significant proportion of humanity, especially through rice harvests in tropical and subtropical Asia. These areas are vulnerable to even a single meter of sea level rise, based on surges in the increasing mega-storms that the second decade of the twenty-first century has experienced to date. These are like platforms from which storm waves move toward areas that are susceptible to both erosion and biological destruction by repeated inundation by salt water during events such as Hurricane Sandy, which so affected the northeastern United States in 2012. The increasing sea level rise prior to A.D. 2300 will intersect and parallel the predicted rise of human population to 9 billion to 10 billion in the last half of the twenty-first century. It is predicted that there will be a slow decline during the first half of the twenty-second century.[4]

Sea level rise will become the single largest financial drain on the world's economies—apart from the world's militaries, that is. The costs will come: from things as simple yet expensive as the cost of raising the many world airports built on landfill next to the sea (Honolulu, San Francisco, Sydney, Hong Kong, Tokyo, etc.) to the less intuitive cost of refitting ship cargo terminals and other infrastructure that would be affected by sea level rise (such as major highways and railroad lines that are more often than not parallel to shorelines in low-altitude flooding caused by storm surge piled upon regional sea level rises). Perhaps paradoxically, sea level rise will in all probability result in an increase in global military spending—to better protect vanishing farmlands from hungry neighbors?

2. *Human mortality caused by an increase in climate events.* The many climate events that seem related to rising global temperatures and greenhouse gas concentrations will exact an ever-greater level of human mortality with every new decade of the twenty-first century. Already we are seeing a quantitative change in terms of human mortality from climate-induced events, with the newest kind of events being the advent of soaking rains of unprecedented volume as well as a well-known human killer: drought. And whereas many of these in the past were longer term (multiyear droughts), it does appear that the incidence of climate events causing human mortality (short-term

violent hurricanes, typhoons, and tornadoes) will be accompanied by equivalents to the Louisiana flooding of 2016 based on a combination of increased rainfall beyond the twentieth-century "normal."

Another new danger will be excessive heat in places where such long periods of summer heat were relatively rare. In places such as Europe, where air-conditioning is rare in private homes and flats, increasing numbers of the aged have been dying during spells of temperatures greater than 40°C (104°F) lasting more than a week. There may be regions on Earth (such as Australia's Outback) that will no longer be habitable by humans because for at least part of the year the temperatures will be too high for humans to tolerate.[5]

While so many focus (and justly so) on high temperatures, perhaps the most consequential event of all was the flooding of Houston by extraordinary rains.[6] This super-rainfall event was quickly forgotten amid the ravages of the hurricane that followed soon after, devastating Puerto Rico as well as other islands in the Caribbean. But it was the Houston rainfall event that was a harbinger of the future. Global warming puts more moisture into the atmosphere. A warmer world increases evaporation of the oceans. Thus, when the inevitable record snowfalls occur, such as happened in Pennsylvania in the last week of the horrid year of 2017, ignorant people, including ignorant politicians, cite the event as "proof" that there is no global warming. If we had a time machine to send these politicians back 100 million years to the Cretaceous Earth, considered the warmest time since the evolution of animals, there would have been snow even then in all probability.

3. *Stress-produced diseases, autism, and depression.* A cornucopia of various toxins affecting human mood and behavior may be already occurring. The increase of toxin levels in our environment is an environmental change, and those in the past have accelerated evolutionary change. But even beyond the chemicals. All of these litanies produce stress, which can cause evolutionary change, given high enough concentration or duration.[7]

4. Regional famines. It is possible that modern agriculture's accelerated crop yield per farmable Earth acre can balance the twin factors of ever more humans and ever less land to grow food for those humans. This goes back to the first warning by Malthus, in Darwin's time.[8]

5. Effects of isolation coming from electronic (non)connectedness. Not enough is yet known of the potential evolutionary effects of increasing isolation amid crowds and brains tied into computers for significant portions of each day. Humanity is conducting a vast experiment. Smartphones are dumbing us down, but even more, they are changing how we think, at least in the amount of time we can ponder a single thought before we are interrupted by a new one.

6. A multiweapon nuclear exchange. Whereas the many works of fiction picturing a twentieth-century nuclear war usually portray a resultant post-apocalyptic landscape, the reality post-2020 is of a "limited" nuclear exchange. India to Pakistan. Iran to Saudi Arabia. North Korea to South Korea, or United States against any of Iran, North Korea, or even Russian or Chinese territories. Especially the possibility of Iran-Israel. Not a thousand bombs, but a half dozen.

This is not an unreal prediction. What the postwar world would be like would depend on the longitude and latitude of the exchange, assuming it would be a short-distance, back-and-forth volley over hundreds of miles, not several thousand. As Carl Sagan put it, this would produce a "nuclear winter." Especially if the United States decides to use nukes to take out the North Korean underground facilities where, supposedly, their nuclear bombs and "new" (aka Russian) missile technology is produced. Bombs built to project great force downward will need to be exploded on ground contact, rather than the air bursts that were used on Hiroshima and Nagasaki and remain the choice of American, Russian, Indian, British, French, Israeli, North Korean, and Pakistani generals as a way to kill as many civilians as possible. The American bombs detonated by ground contact (bunker busters!) will create far longer and more devastating climate effects than airbursts because they will throw much more sediment into the air.

7. *We somehow avoid all of these futures and through effort, intelligence, and goodwill (to say nothing of necessity) and actually live as a peaceful species.* Optimism—that technology such as fusion energy sources, or a real commitment to non-carbon-driven energy sources, that the CRISPR-Cas9 and its equivalents do conquer disease and lead to more food. And so on. That our better angels do prevail.

Yes, there is a possibility that the human affairs even as recently as 2016 have increased stress in many people, and in so doing have perhaps contributed to human evolution. One symptom is a quantifiable change in behavior at least in America: the increased diet of news channels consumed per day, and done along political party lines. Another is the radically increased number of shootings of innocents in America from 2016 on.

In spring of 2017, I received an e-mail from the website *Gizmodo*,[9] asking if I would comment on the evolutionary possibilities of genetically producing "supersoldiers." I agreed to a phone interview and talked to a reporter for the site for perhaps thirty minutes, and then I promptly forgot the whole thing. But not for long. Soon after the interviews (there were a number of academics interviewed as I was) were published on the site, I was bombarded by e-mails coming from both the interviewer and the numerous sites that then picked up on a statement that I do not even remember making: It was claimed that I stated that the election and by then six-month-old presidency of Donald Trump was causing evolutionary change in humanity, and not in any positive direction. The backlash was monumental, starting with Fox News screeching about such treason, to letters to the president and board of regents of the University of Washington, to major evolutionists taking to their Twitter accounts and firing all their rhetorical, dismissive broadsides at this nonsensical notion.[10] One such refrain was that I was "politicizing" science. As if that had never happened before. It was this that made me decide to make Michael Skinner one of the people to whom this book is dedicated. His own 2005 assertion of the evolutionary dangers of farmland chemicals led to his crucifixion online organized by big-money chemical companies.

I maintain that, in 2017 and now in 2018 as I write this, there continues to be an increase in stress compared to, perhaps, a decade or even a half decade ago, among average Americans and Europeans, at least. As one bit of evidence, the increase in the number of the truly evil insane—the mass shooters—can be seen in the diagram on the following page, previously unpublished.

My statement to *Gizmodo* was wedded in the same arguments that make up the crux of this book: that certain environmental factors, be they physical or social, can conceivably create sufficient increases in global human stress levels that some epigenetic changes might occur. Perhaps a single day of combat, but perhaps six months of elevated stress. No one knows yet. But my working hypothesis is that for all of 2016–17 and going into 2018, certainly in America but probably in many other parts of the globe as well, many humans are in indeed living in a heightened sense of stress brought about by a rapidly changing social and physical (environmental) worldscape. The graph of the mass shootings shown here should be evidence enough that we are in a new period in human history, one in which we are combining the more archaic savagery of the Middle Ages with social media and an advanced and connected computer age of smartphones, which themselves might be evolving us. When coupling smartphones with other sources of increased stress (since smartphones are stress generators in many people), it will be interesting to look at the DNA in humanity in a generation, and also at the epigenome.

So, a thought experiment goes back to one that is similar to what University of Adelaide biologists[11] have conducted on the bones of late Ice Age mammals. They excavated and then analyzed for epigenetic

MASS SHOOTINGS IN AMERICA

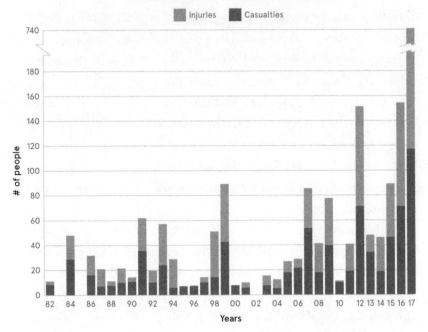

Mass shooting injuries and deaths in the U.S. have increased greatly in recent years, a significant source of stress that may cause evolution through increased cortisol production. Data set created and maintained by Mark Follman and *Mother Jones*, https://www.motherjones.com/politics/2012/12/mass-shootings -mother-jones-full-data/.

markers in fossils bones of animals living in East Asia (now Siberia) prior to human contact, and compared the results from pre–human contact mammals to results from the bones of younger fossils coming from prey species that lived among the human hunters for the first time. In the case of pre- and post- human food sources, they compared the number of stress-related epigenetic marks found in the bones of late Ice Age musk ox and other large herbivores. These human invaders were armed for the first time with large flint spear tips on spears capable of killing large Ice Age mammals. As these hunters moved eastward through Asia, some tens of thousands of years ago, their appearance coincided with a measureable increase in stress levels as evidenced from this chemical study of their bones. Just as surely, this major environmental change had to have caused behavioral change in these animals that had heretofore not been hunted by humans.

So the thought experiment is whether Native Americans' stress increased after the appearance of European invaders. Such an experiment on human bones will hopefully never happen, because we would be desecrating the long dead. But I suspect the first appearance of Europeans across North and South America caused major stress in the Native Americans.

We all have heard narratives about animals that lived in habitats that had excluded humans, and the docility of these animals on first contact with humans, who were not at first considered dangerous to these Garden of Eden analogues. Relative stress levels among Native Americans living in the North American coastal Southeast were no doubt comparable before and some decades after the arrival of Europeans.

Over the years of writing this book (begun in 2014), I made numerous queries of practicing psychologists, asking a similar question about our own time: Did they think that their patients showed a real increase in stress-related problems, or even in perception of their stress levels, compared to even five years earlier? It was a small sample, but in every case it seemed that stress has vastly increased in the past five years. A more quantitative finding is what is called the "opiate crisis": that a higher percentage of the American population is using opiates than at any time in at least the twentieth century. Add to that the levels of meth-amphetamine, marijuana, and alcohol abuse, and we see other evidence of the American human population undergoing major behavioral change.

The final aspect of this short narrative was that in some of the follow-up questions I received by various reporters asking if I *really* believed that the presidency of Donald Trump would have evolutionary consequences, I answered that I believe that stress levels would be just as high had Hillary Clinton won. And, yes, I think there are potential evolutionary consequences to this time, not least of which is that the human murder rate (a part of violent crime)—which, as noted in a previous chapter, has been on the rise for the past several years after more than twenty years of decline—has not peaked. That peak of murder rate I predict to occur around 2020, followed by decline.

If all Americans could magically have their serum blood levels assessed for the quantifies of various stress molecules in them, as well as the more joyous and or "contentment" molecules, such as serotonin, what would be the mean values by age, gender, race, religion, and wealth?

INTO THE FUTURE

One of the goals of these pages has been to try to report on two quite different potential processes of evolutionary change: how rapid physical environmental change among the living organisms of planet Earth in the "deep time" of life's history as well as far more recent environmental changes in *human* history have both been major drivers of the history of life on Earth. The evolutionary consequences accompanying the vast "environmental" changes during the origin, rise, and fall of human civilizations may have been not so different from the asteroid impacts or volcanic outpourings in causing evolutionary change. For us humans, however, that evolution might have been more concentrated in new kinds of behavior, rather than new kinds of biological body plans.

At present, the greatest critics of heritable epigenetics are found at the pinnacle of their academic careers. Often their statements can be summarized as a series of "No. No. No. IMPOSSIBLE!" To them, I offer this quote from the great Arthur C. Clarke: "If an elderly but distinguished scientist says that something is possible he is almost certainly right, but if he says that it is impossible he is very probably wrong."[12]

Is it possible that epigenetic processes have been extremely important in the story of Life on Earth, and also the arc of human evolution and behavior?

Notes

PREFACE: THE JURASSIC PARK OF NEVADA

1. Regarding "insensibly graded series," there are many who describe this view that Darwinian theory demands that all transitional forms should be found in the fossil record. One of many such references is in Richard Dawkins, *The Blind Watchmaker: Why the Evidence of Evolution Reveals a Universe Without Design* (New York: W. W. Norton & Company, 2015).

2. Peter Ward and Joe Kirschvink, *A New History of Life: The Radical New Discoveries About the Origins and Evolution of Life on Earth* (New York: Bloomsbury Press, 2015).

3. Ibid.

4. *Wikipedia*, s.v. "Modern Synthesis (20th Century)," last modified January 27, 2018, en.wikipedia.org/wiki/Modern_synthesis_(20th_century).

5. Thomas S. Kuhn, *The Structure of Scientific Revolutions*, 2nd ed. (Chicago: University of Chicago Press, 1970).

6. For a further definition of *epigenetics*, a great place to start is Nessa Carey, *The Epigenetics Revolution: How Modern Biology Is Rewriting Our Understanding of Genetics, Disease, and Inheritance* (New York: Columbia University Press, 2012).

INTRODUCTION: LOOKING BACK

1. Jacqueline Howard, "Americans Devote More Than 10 Hours a Day to Screen Time, and Growing," CNN, updated July 29, 2016, https://www.cnn.com/2016/06/30/health/americans-screen-time-nielsen/index.html.

2. Theodosius Dobzhansky, "Nothing in Biology Makes Sense Except in the Light of Evolution," *American Biology Teacher* 35, no. 3 (March 1973): 125–29.

3. For more on lateral gene transfer, see Howard Ochman, Jeffrey G. Lawrence, and Eduardo A. Groisman, "Lateral Gene Transfer and the Nature of Bacterial Innovation," *Nature* 405, no. 6784 (May 2000): 299–304, and J. C. Dunning Hotopp, "Horizontal Gene Transfer Between Bacteria and Animals," *Trends in Genetics*. 27, no. 4 (April 2011): 157–63.

4. Eva Jablonka and Marion J. Lamb, *Evolution in Four Dimensions: Genetic, Epigenetic, Behavioral, and Symbolic Variation in the History of Life* (Cambridge, MA: MIT Press, 2005).

5. There are many references to this. It is certainly *not* the Sixth Extinction, as Kirschvink and I pointed out. It is at least the tenth *major* mass extinction—if it is yet one at all! One of the earliest books on this was my own, Peter Ward, *The End of Evolution: On Mass Extinctions and the Preservation of Biodiversity* (New York: Bantam, 1994), and more recently Elizabeth Kolbert, *The Sixth Extinction: An Unnatural History* (New York: Henry Holt, 2014).

6. On stress and rat parenting, see Genetic Science Learning Center, "Lick Your Rats," learn.genetics.utah.edu/content/epigenetics/rats; stress and social environment: Kathryn Gudsnuk and Frances A. Champagne, "Epigenetic Influence of Stress and the Social Environment," *ILAR Journal* 53, no. 3–4 (December 2012): 279–88; Gudsnuk and Champagne, "Epigenetic Effects of Early Developmental Experiences," *Clinics in Perinatology* 38, no. 4 (December 2011): 703–17; Champagne, "Epigenetic Influence of Social Experiences Across the Lifespan," *Developmental Psychobiology* 52, no. 4 (2010): 299–311. Also see Andreas von Bubnoff, "Does Stress Speed Up Evolution?" *Nautilus*, March 31, 2016, nautil .us/issue/34/adaptation/does-stress-speed-up-evolution.

7. A thought experiment that I thought I thought up first! No. Dawn Papple, "Epigenetic Inheritance: Holocaust Study Proves What Native Americans Have 'Always Known,'" *Inquisitr*, August 24, 2015, inquisitr.com/2352952/epigenetic-inheritance -holocaust-study-proves-what-native-americans-have-always-known.

CHAPTER I: FROM GOD TO SCIENCE

1. Gregory Bateson, *Steps to an Ecology of Mind: Collected Essays in Anthropology, Psychiatry, Evolution, and Epistemology* (Chicago: University of Chicago Press, 1972), 259.

2. Jean-Baptiste Lamarck, *Philosophie zoologique; ou, Exposition des considérations relatives à l'histoire naturelle des animaux* (Dentu: Paris, 1809).

3. Konrad Guenther, *Darwinism and the Problems of Life; a Study of Familiar Animal Life* (London: A. Owen, 1906); Eva Jablonka and Marion J. Lamb, "The Transformations of Darwinism," chap. 1 in *Evolution in Four Dimensions: Genetic, Epigenetic, Behavioral, and Symbolic Variation in the History of Life* (Cambridge, MA: MIT Press, 2005).

4. Thomas S. Kuhn, *The Structure of Scientific Revolutions*, 2nd ed. (Chicago: University of Chicago Press, 1970).

5. Ibid.

6. Theodosius Dobzhansky, "Nothing in Biology Makes Sense Except in the Light of Evolution," *American Biology Teacher* 35, no. 3 (March 1973): 125–29; Ernst Mayr, *The Growth of Biological Thought: Diversity, Evolution, and Inheritance* (Cambridge, MA: Belknap Press, 1985).

7. From Jerry Coyne, the leading critic of epigenetics: "Is 'Epigenetics' a Revolution in Evolution?" *Why Evolution Is True* (blog), whyevolutionistrue.wordpress

.com/2011/08/21/is-epigenetics-a-revolution-in-evolution. Even more recently, the outcry about epigenetics following an article in the *New Yorker* (Siddhartha Mukherjee, "Same but Different," May 2, 2016) contained an astonishing cascade of bile, as recounted here: Tabitha M. Powledge, "That Mukherjee Piece on Epigenetics in the *New Yorker*," *On Science Blogs*, May 13, 2016, blogs.plos.org /onscienceblogs/2016/05/13/that-mukherjee-piece-on-epigenetics-in-the -new-yorker; see, for example: Mark Ptashne and John Greally, "Researchers Criticize the Mukherjee Piece on Epigenetics: Part 2," *Why Evolution Is True* (blog), May 6, 2016, whyevolutionistrue.wordpress.com/2016/05/06/researchers -criticize-the-mukherjee-piece-on-epigenetics-part-2.

8. Anthony M. Alioto, *A History of Western Science* (Englewood Cliffs, NJ: Prentice Hall, 1987); David C. Lindberg, *The Beginnings of Western Science: The European Scientific Tradition in Philosophical, Religious, and Institutional Context, 600 B.C. to A.D. 1450* (Chicago: University of Chicago Press, 1992).

9. Frank Dawson Adams, *The Birth and Development of the Geological Sciences* (Baltimore: Williams & Wilkins, 1938); Peter J. Bowler, *The Earth Encompassed: A History of the Environmental Sciences* (New York: Norton, 2000).

10. James Ussher, *The Annals of the World* (1650); James Barr, "Why the World Was Created in 4004 B.C.: Archbishop Ussher and Biblical Chronology," *Bulletin of the John Rylands* 67 (1984–85): 575–608.

11. By far the author who has written best about how fossils were first identified and then used is Martin J. S. Rudwick, particularly his classic *The Meaning of Fossils: Episodes in the History of Palaeontology* (New York: American Elsevier, 1972), but also his more recent books *The New Science of Geology: Studies in the Earth Sciences in the Age of Revolution* (Burlington, VT: Ashgate, 2004); *Lyell and Darwin, Geologists: Studies in the Earth Sciences in the Age of Reform* (Burlington, VT: Ashgate, 2005); and *Earth's Deep History: How It Was Discovered and Why It Matters* (Chicago: University of Chicago Press, 2014).

12. Rudwick, *The Meaning of Fossils*.

13. Ibid.

14. Ibid.

15. Margaret J. Anderson, *Carl Linnaeus: Father of Classification* (Springfield, NJ: Enslow, 1997).

16. Ibid.

17. Mayr, *The Growth of Biological Thought*, 330; Otis E. Fellows and Stephen F. Milliken, *Buffon* (New York: Twayne, 1972), 149–54.

18. Georges-Louis Leclerc (Comte de Buffon), *L'Histoire naturelle, générale et particuliére, avec la description du Cabinet du Roi* (Paris: Imprimerie Royale, 1789).

19. Erasmus Darwin, *Phytologia; or, The Philosophy of Agriculture and Gardening* (London: J. Johnson, 1800), 77; Patricia Fara, Erasmus Darwin: *Sex, Science, and Serendipity* (Oxford, England: Oxford University Press, 2012).

20. Erasmus Darwin, *Zoonomia; or, the Laws of Organic Life* (London: Thomas and Andrews), 1:397.

21. Ibid.; Stephen Foster, "The Decline in Erasmus Darwin's Reputation and His Legacy," *Victorian Web* (blog), November 5, 2016, victorianweb.org/science /edarwin/reputation.html.

CHAPTER II: LAMARCK TO DARWIN

1. Charles Darwin, *On the Origin of Species* (London: John Murray, 1859); Richard W. Burkhardt Jr. "Lamarck, Evolution, and the Politics of Science," *Journal of the History of Biology* 3, no. 2 (September 1970): 275–98; William Coleman, *Biology in the Nineteenth Century: Problems of Form, Function, and Transformation* (Cambridge: Cambridge University Press, 1977).

2. Richard W. Burkhardt Jr. "Lamarck, Evolution, and the Politics of Science," *Journal of the History of Biology* 3, no. 2 (September 1970): 275–98; William Coleman, *Biology in the Nineteenth Century: Problems of Form, Function, and Transformation* (Cambridge: Cambridge University Press, 1977).

3. Ernst Mayr, *The Growth of Biological Thought: Diversity, Evolution, and Inheritance* (Cambridge, MA: Belknap Press, 1985), 356.

4. Georges Cuvier, "Elegy of Lamarck," *Edinburgh New Philosophical Journal* 20 (January 1836): 1–22.

5. Martin J. S. Rudwick, *Georges Cuvier, Fossil Bones, and Geological Catastrophes: New Translations and Interpretations of the Primary Texts* (Chicago: University of Chicago Press, 1998).

6. Stephen Jay Gould, *The Structure of Evolutionary Theory* (Cambridge, MA: Belknap Press, 2002); Ross Honeywill, *Lamarck's Evolution: Two Centuries of Genius and Jealousy* (Sydney: Pier 9, 2008).

7. Martin J. S. Rudwick, *The Meaning of Fossils: Episodes in the History of Paleontology* (New York: American Elsevier, 1972); Rudwick, *Georges Cuvier.*

8. There is a good summary, titled "Extinctions: Georges Cuvier," on the invaluable website from the University of California, Berkeley, *Understanding Evolution*, evolution.berkeley.edu/evolibrary/article/history_08; see also Elizabeth Kolbert's great read from the *New Yorker*, "The Lost World," December 16, 2013.

9. Georges Cuvier, *Note on the Species of Living and Fossil Elephants* (Paris: n.p., 1796).

10. Darwin, "On the Geological Succession of Organic Beings: On Extinction," chap. 10 in *On the Origin of Species.*

11. Rudwick, *Georges Cuvier.*

12. Quoted in Gould, *Structure of Evolutionary Theory*, 491.

13. Rudwick, *The Meaning of Fossils*; Rudwick, *Georges Cuvier.*

14. Ibid.

15. On catastrophism, see Alexander H. Taylor, "Catastrophism," *Foundation of Modern Geology* (blog), publish.illinois.edu/foundationofmoderngeology/cata strophism). The theory was accepted in Cuvier's time, then rejected in the twentieth century, but now has been resurrected in this century as "neo-catastrophism," aided in no small way by the discovery that an asteroid was a prime cause of the extinction of dinosaurs. See Trevor Palmer, *Catastrophism, Neocatastrophism and Evolution* (Nottingham, England: Society for Interdisciplinary Studies/Nottingham Trent University, 1994).

16. Alpheus S. Packard, *Lamarck, the Founder of Evolution: His Life and Work* (New York: Longmans, Green, 1901); Jean-Baptiste Lamarck, *Philosophie zoologique; ou, Exposition des considérations relatives à l'histoire naturelle des animaux* (Paris: Dentu, 1809).

17. Jean-Baptiste Lamarck, *Encyclopédie méthodique: Botanique*, 8 vols. and suppl. (Paris: Panckoucke, 1783–1817). Then: Jean-Baptiste Lamarck, *Système des animaux sans vertèbres; ou, Tableau général des classes, des ordres et des genres de ces animaux . . .* (Paris: Detreville, 1801) VIII: 1–432.1815–22; Jean-Baptiste Lamarck *Histoire naturelle des animaux sans vertèbres . . .* , 7 vols. (Paris: Verdière, 1815–22).

18. Ibid.

19. For a good recent summary of the new look at Lamarckism, see Emily Singer, "A Comeback for Lamarckian Evolution?" *MIT Technology Review*, February 4, 2009, www.technologyreview.com/s/411880/a-comeback-for-lamarckian-evo lution. Also, for a great summary of how modern biology continues to misquote and denigrate Lamarck and his work, see Michael T. Ghiselin, "The Imaginary Lamarck: A Look at Bogus 'History' in Schoolbooks," *Textbook Letter*, September–October 1994, textbookleague.org/54marck.htm. And finally: Eva Jablonka and Marion J. Lamb, *Evolution in Four Dimensions: Genetic, Epigenetic, Behavioral, and Symbolic Variation in the History of Life* (Cambridge, MA: MIT Press, 2005). See also: Dan Graur, Manolo Gouy, and David Wool, "In Retrospect: Lamarck's Treatise at 200," *Nature* 460, no. 7256 (August 2009): 688–89; Richard W. Burkhardt Jr., *The Spirit of System: Lamarck and Evolutionary Biology* (Cambridge, MA: Harvard University Press, 1995).

CHAPTER III: FROM DARWIN TO THE NEW (MODERN) SYNTHESIS

1. David Lack, "Evolution of the Galapagos Finches," *Nature* 146, no. 3697 (September 1940): 324–27. But irony, irony, irony! See the astonishing new work co-authored by Michael Skinner, to whom this book is codedicated: Sabrina M. McNew et al., "Epigenetic Variation Between Urban and Rural Populations of Darwin's Finches," *BMC Evolutionary Biology* 17, no. 1 (2017): doi.org/10.1186 /s12862-017-1025-9.

2. Charles Darwin, *The Variation of Animals and Plants Under Domestication*, 2 vols. (London: John Murray, 1868).

3. Thomas Malthus, *An Essay on the Principle of Population . . .* (London: J. Johnson, 1798); see also Adrian Desmond and James Moore, *Darwin* (London: Penguin, 1992).

4. Malthus, *An Essay on the Principle of Population*, 44.

5. My late and greatly missed friend Steve Gould examines this concept in great detail in the ultimate book of his long career: Stephen Jay Gould, *The Structure of Evolutionary Theory* (Cambridge, MA: Belknap Press, 2002).

6. Ibid.

7. Richard Burkhardt Jr. "Lamarck, Evolution and Inheritance of Acquired Characters," *Genetics* 194 (2013): 793–805.

8. From a lecture delivered by Lamarck to the Muséum National d'Histoire Naturelle, Paris, May 1803.

9. Desmond and Moore, *Darwin*.

10. Charles Doolittle Walcott, "Searching for the First Forms of Life," lecture, c. 1892–1894, quoted in Stephen Jay Gould, *Wonderful Life: The Burgess Shale and the Nature of History* (New York: W. W. Norton Co., 1989).

11. Charles Darwin, *On the Origin of Species* (London: John Murray, 1859).

12. Darwin, "Recapitulation and Conclusion: Recapitulation," chap. 14 in *On the Origin of Species*.

13. George Gaylord Simpson, *The Major Features of Evolution* (New York: Columbia University Press, 1953), 360.

14. George Gaylord Simpson, *Tempo and Mode in Evolution*, rev. ed. (New York: Columbia University Press, 1984); George Gaylord Simpson, *The Meaning of Evolution* (New York: Mentor, 1951). See also Jay D. Aronson, "'Molecules and Monkeys': George Gaylord Simpson and the Challenge of Molecular Evolution," *History and Philosophy of the Life Sciences* 24, nos. 3–4 (2002): 441–65, as well as the great work of my friend Léo Laporte: Léo F. Laporte, "Simpson on Species," *Journal of the History of Biology* 27, no. 1 (March 1994): 141–59.

15. Frank Fenner and I. D. Marshall, "A Comparison of the Virulence for European rabbits (*Oryctolagus cuniculus*) of Strains of Myxoma Virus Recovered in the Field in Australia, Europe and America," *Journal of Hygiene* 55, no. 2 (June 1957): 149–91.

16. On allopatric speciation, see Nelson R. Cabej, "Species and Allopatric Speciation," in *Epigenetic Principles of Evolution* (Waltham, MA: Elsevier, 2012), 707–23.

17. For the emerging uncertainty that began in the twenty-first century, see Eugene V. Koonin, "The *Origin* at 150: Is a New Evolutionary Synthesis in Sight?" *Trends in Genetics* 25, no. 11 (November 2009): 473–75; and Eugene V. Koonin, *The Logic of Chance: The Nature and Origin of Biological Evolution* (Upper Saddle River,

NJ: FT Press, 2011). For more on allopatric speciation, see Jerry A. Coyne and H. Allen Orr, *Speciation* (Sunderland, MA: Sinauer Associates, 2004), 83–124; Michael Turelli, Nicholas H. Barton, and Jerry A. Coyne "Theory and Speciation," *Trends in Ecology & Evolution* 16 no. 7 (August 2001): 330–43; H. Allen Orr and Lynne H. Orr "Waiting for Speciation: The Effect of Population Subdivision on the Time to Speciation," *Evolution* 50, no. 5 (October 1996): 1742–49.

18. Walter Sullivan, "Luis W. Alvarez, Nobel Physicist Who Explored Atom, Dies at 77," *New York Times*, September 2, 1988.

19. The amazing thing noted by many other authors was that the original scientific definition of *punctuated equilibria* by Eldredge and Gould was published in a book that would have been seen only by paleontologists: Niles Eldredge and Stephen Jay Gould, "Punctuated Equilibria: An Alternative to Phyletic Gradualism," in *Models in Paleobiology*, ed. Thomas J. M. Schopf (San Francisco: Freeman Cooper, 1972), 82–115. Soon after, a new version came out: Stephen Jay Gould and Niles Eldredge, "Punctuated Equilibria: The Tempo and Mode of Evolution Reconsidered," *Paleobiology* 3, no. 2 (Spring 1977): 115–51. It was later reprinted in Niles Eldredge, *Time Frames: The Rethinking of Darwinian Evolution and the Theory of Punctuated Equilibria* (New York: Simon & Schuster, 1985), 193–223. But such was the power of this hypothesis that it transcended the field of paleontology, spilling out into evolutionary fields and having other kinds of influence as well: Francisco J. Ayala, "The Structure of Evolutionary Theory: On Stephen Jay Gould's Monumental Masterpiece," *Theology and Science* 3, no. 1 (2005): 104.

20. Gould and Eldredge, "Punctuated Equilibria."

21. Stephen Jay Gould, "Evolution's Erratic Pace," *Natural History* 86 (May 1977).

22. Koonin, "The *Origin* at 150"; and Koonin, *The Logic of Chance.*

23. A good introduction to the views of Eva Jablonka, who has been the most influential proponent of "soft inheritance," is Laurence A. Moran, "Extending Evolutionary Theory?—Eva Jablonka," *Sandwalk* (blog), October 2, 2016, sandwalk.blogspot.com/2016/10/extending-evolutionary-theory-eva.html; and here is one of the most influential of all the summaries of how the new field of epigenetics is impacting the theory of evolution: Eugene V. Koonin and Yuri I. Wolf, "Is Evolution Darwinian or/and Lamarckian?" *Biology Direct* 4, no. 42 (2009): oi .org/10.1186/1745-6150-4-42.

CHAPTER IV: EPIGENETICS AND THE NEWER SYNTHESIS

1. Bruce Saunders and Peter Ward, "Sympatric Occurrence of Living Nautilus (*N. Pompilius* and *N. Stenomphalus*) on the Great Barrier Reef, Australia," *Nautilus* 101, no. 4 (1987): 188–93.

2. Lauren E. Vandepas, Frederick D. Dooley, Gregory J. Barord, Billie J. Swalla, and Peter D. Ward, "A Revisited Phylogeography of *Nautilus Pompilius*," *Ecology and Evolution* 6, no. 14 (July 2016): 4924–35.

3. Thomas H. Clarke Jr., "The Columbian and Woolly Mammoth May Be One Highly Variable Species," *LexisNexis Legal Newsroom: Environmental* (blog), January 21, 2012, lexisnexis.com/legalnewsroom/environmental/b/fishwildlife /archive/2012/01/21/the-columbian-and-woolly-mammoth-may-be-one-highly -variable-species.aspx?.

4. Eva Jablonka and Marion J. Lamb, *Evolution in Four Dimensions: Genetic, Epigenetic, Behavioral, and Symbolic Variation in the History of Life* (Cambridge, MA: MIT Press, 2005); Eva Jablonka and Marion J. Lamb, "Epigenetic Inheritance in Evolution," *Journal of Evolutionary Biology* 11, no. 2 (March 1998): 159–83.

5. Eugene V. Koonin and Yuri I. Wolf, "Is Evolution Darwinian or/and Lamarckian?" *Biology Direct* 4, no. 42 (2009): oi.org/10.1186/1745-6150-4-42.; see also Eva Jablonka, "Epigenetic Inheritance and Plasticity: The Responsive Germline," *Progress in Biophysics and Molecular Biology* 111, no. 2–3 (April 2013): 99–107.

6. C. H. Waddington, "The Basic Ideas of Biology," in *Towards a Theoretical Biology: Prolegomena*, ed. C. H. Waddington (Edinburgh: Edinburgh University Press, 1968), 1–32.

7. Michael Turelli, Nicholas H. Barton, and Jerry A. Coyne "Theory and Speciation," *Trends in Ecology & Evolution* 16, no. 7 (August 2001): 330–43. Contrast this with Catherine M. Suter, Dario Boffelli, and David I. K. Martin, "A Role for Epigenetic Inheritance in Modern Evolutionary Theory? A Comment in Response to Dickins and Rahman," *Proceedings of the Royal Society B: Biological Sciences* 280, no. 1771 (November 2013): doi: 10.1098/rspb.2013.0903.

8. Genetic Science Learning Center, "Lick Your Rats," learn.genetics.utah.edu /content/epigenetics/rats.

9. Ian C. G. Weaver et al., "Epigenetic Programming by Maternal Behavior," *Nature Neuroscience* 7 (2004): 847–54.

10. Elizabeth J. Duncan, Peter D. Gluckman, and Peter K. Dearden, "Epigenetics, Plasticity and Evolution: How Do We Link Epigenetic Change to Phenotype?" *Journal of Experimental Zoology Part B: Molecular and Developmental Evolution* 322, no. 4 (June 2014): 208–20; Rodrigo S. Galhardo, P. J. Hastings, and Susan M. Rosenberg, "Mutation as a Stress Response and the Regulation of Evolvability," *Critical Reviews in Biochemistry and Molecular Biology* 42, no. 5 (2007): 399–435; Susan M. Rosenberg, "Mutation for Survival," *Current Opinion in Genetics and Development* 7, no. 6 (December 1997): 829–34.

11. Dan Graur, Manolo Gouy, and David Wool, "In Retrospect: Lamarck's Treatise at 200," *Nature* 460, no. 7256 (August 2009): 688–89; Richard W. Burkhardt Jr., *The Spirit of System: Lamarck and Evolutionary Biology* (Cambridge, MA: Harvard University Press, 1995).

12. Francis Crick, "Central Dogma of Molecular Biology," *Nature* 227, no. 5258 (August 1970): 561–63.

13. Nessa Carey, *The Epigenetics Revolution: How Modern Biology Is Rewriting Our Understanding of Genetics, Disease, and Inheritance* (New York: Columbia University Press, 2012).

14. Ibid.

15. Koonin and Wolf, "Is Evolution Darwinian or/and Lamarckian?"

16. Jean Gayon, "From Mendel to Epigenetics: History of Genetics," *Comptes Rendus Biologies* 339, no. 7–8 (August 2016): 225–30.

17. Sander Gliboff, "'Protoplasm . . . Is Soft Wax in Our Hands': Paul Kammerer and the Art of Biological Transformation," *Endeavour* 29, no. 4 (December 2005): 162–67; Alexander O. Vargas, "Did Paul Kammerer Discover Epigenetic Inheritance? A Modern Look at the Controversial Midwife Toad Experiments," *Journal of Experimental Zoology Part B: Molecular and Developmental Evolution* 312B, no. 7 (November 2009): 667–78.

18. Zhores A. Medvedev, *The Rise and Fall of T. D. Lysenko*, trans. I. Michael Lerner (New York: Columbia University Press, 1969).

19. Koonin and Wolf, "Is Evolution Darwinian or/and Lamarckian?"

20. Jean-Baptiste Lamarck, *Philosophie zoologique; ou, Exposition des considérations relatives à l'histoire naturelle des animaux* (Paris: Dentu, 1809).

21. Mark A. Rothstein, Yu Cai, and Gary E. Merchant, "The Ghost in Our Genes: Legal and Ethical Implications of Epigenetics," *Health Matrix* 19, no. 1 (Winter 2009): 1–62.

22. John van der Oost et al., "CRISPR-Based Adaptive and Heritable Immunity in Prokaryotes," *Trends in Biochemical Sciences* 34, no. 8 (2009): 401–7.

23. Gene W. Tyson and Jillian F. Banfield, "Rapidly Evolving CRISPRs Implicated in Acquired Resistance of Microorganisms to Viruses," *Environmental Microbiology* 10, no. 1 (January 2008): 200–7.

24. Rotem Sorek, Victor Kunin, and Philip Hugenholtz, "CRISPR—a Widespread System That Provides Acquired Resistance Against Phages in Bacteria and Archaea," *Nature Reviews Microbiology* 6, no. 3 (March 2008): 181–86.

25. Alex Reis et al., "CRISPR/Cas9 and Targeted Genome Editing: A New Era in Molecular Biology," New England BioLabs, www.neb.com/tools-and-resources /feature-articles/crispr-cas9-and-targeted-genome-editing-a-new-era-in -molecular-biology.

26. Ibid.

27. Jablonka and Lamb, *Evolution in Four Dimensions*; Jablonka and Lamb, "Epigenetic Inheritance in Evolution," *Journal of Evolutionary Biology* 11, no. 2 (March 1998): 159–83.

28. Rebecca M. Terns and Michael P. Terns, "CRISPR-Based Technologies: Prokaryotic Defense Weapons Repurposed," *Trends in Genetics* 30, no. 3 (March 2014): 111–18; Eugene V. Koonin and Yuri I. Wolf, "Genomics of Bacteria and Archaea: The Emerging Dynamic View of the Prokaryotic World," *Nucleic Acids Research* 36, no. 21 (December 2008): 6688–719.

29. Ewen Callaway, "Fearful Memories Haunt Mouse Descendants," *Nature* News, December 1, 2013, nature.com/news/fearful-memories-haunt-mouse-descen dants-1.14272.

30. Paul Kammerer, *The Inheritance of Acquired Characteristics* (New York: Boni & Liveright, 1924).

31. Nathaniel Scharping, "How a Russian Scientist Bred the First Domesticated Foxes," *Discover*, September 14, 2016.

32. Francesca Pacchierotti and Marcello Spanò, "Environmental Impact on DNA Methylation in the Germline: State of the Art and Gaps of Knowledge," *Biomed Research International* (2015).

CHAPTER V: THE BEST OF TIMES, THE WORST OF TIMES—IN DEEP TIME

1. On *Tiktaalik*, see Edward B. Daeschler, Neil H. Shubin, and Farish A. Jenkins Jr., "A Devonian Tetrapod-Like Fish and the Evolution of the Tetrapod Body Plan," *Nature* 440, no. 7085 (April 2006): 757–63.II.

2. This "estimate" of how much faster epigenetic change can be than strict Darwinian is just that: an estimate. But the rates cited in various articles are indeed astonishing. See Jenny Rood, "Estimating Epigenetic Mutation Rates," *Scientist*, May 11, 2005, the-scientist.com/?articles.view/articleNo/42948/title /Estimating-Epigenetic-Mutation-Rates; "Researchers Obtain Precise Esti mates of the Epigenetic Mutation Rate," *Phys.org*, May 11, 2015, phys.org/news /2015-05-precise-epigenetic-mutation.html.

3. Yuan-Ye Zhang et al., "Epigenetic Variation Creates Potential for Evolution of Plant Phenotypic Plasticity," *New Phytologist* 197, no. 1 (January 2013): 314–22.; Daniel Nätt et al., "Heritable Genome-Wide Variation of Gene Expression and Promoter Methylation Between Wild and Domesticated Chickens," *BMC Genomics* 13 (2012): doi: 10.1186/1471-2164-13-59.

4. Rates of evolution have been a major topic in paleobiology and evolutionary theory for generations. With the new understanding of epigenetics much of these topics need reconsideration. The whole "gap" question is being looked at anew: William B. Miller Jr., "What Is the Big Deal About Evolutionary Gaps?" in *The Microcosm Within: Evolution and Extinction in the Hologenome* (Boca Raton, FL: Universal-Publishers, 2013), 177, 395–96; "Fastest Evolving Creature Is 'Living Dinosaur,'" *Live Science*, March 25, 2008, livescience.com/2396-fastest -evolving-creature-living-dinosaur.html.

5. All of this discussed at length in Peter Ward and Joe Kirschvink, *A New History of Life: The Radical New Discoveries About the Origins and Evolution of Life on Earth* (New York: Bloomsbury Press, 2015).

6. Vincent Courtillot, "True Polar Wander," in *Encyclopedia of Geomagnetism and Paleomagnetism*, ed. David Gubbins and Emilio Herrero-Bervera (2007

edition), link.springer.com/referenceworkentry/10.1007%2F978-1-4020-4423
-6_308.

7. Joseph L. Kirschvink, "Late Proterozoic Low-Latitude Global Glaciation: The
Snowball Earth," in *The Proterozoic Biosphere: A Multidisciplinary Study*, ed.
J. William Schopf and Cornelius Klein (Cambridge: Cambridge University Press,
1992), 51–52; Frank A. Corsetti, Stanley M. Awramik, and David Pierce, "A
Complex Microbiota from Snowball Earth Times: Microfossils from the Neopro-
terozoic Kingston Peak Formation, Death Valley, USA," *Proceedings of the
National Academy of Sciences of the United States of America* 100, no. 8
(April 2003): 4399–4404; Paul F. Hoffman and Daniel P. Schrag, "The Snowball Earth Hypothesis: Testing the Limits of Global Change," *Terra Nova* 14,
no. 3 (June 2002): 129–55.

8. Tjeerd H. van Andel, *New Views on an Old Planet: A History of Global
Change*, 2nd ed. (Cambridge: Cambridge University Press, 1994); see also a
highly readable article in *New Scientist*: Michael Marshall, "The History of Ice on
Earth," May 24, 2010, www.newscientist.com/article/dn18949-the-history-of-ice
-on-earth.

CHAPTER VI: EPIGENETICS AND THE ORIGIN
AND DIVERSIFICATION OF LIFE

1. Peter Ward, *Life as We Do Not Know It: The NASA Search for (and Synthesis
of) Alien Life* (New York: Viking, 2005).

2. W. Ford Doolittle, "Lateral Genomics," *Trends in Cell Biology* 9, no. 12
(December 1999): M5–8; W. Ford Doolittle and Olga Zhaxybayeva, "On the
Origin of Prokaryotic Species," *Genome Research* 19, no. 5 (May 2009): 744–
56; Eugene V. Koonin, Kira S. Makarova, and L. Aravind, "Horizontal Gene
Transfer in Prokaryotes: Quantification and Classification," *Annual Review of
Microbiology* 55 (October 2001): 709–42; J. Peter Gogarten and Jeffrey P.
Townsend, "Horizontal Gene Transfer, Genome Innovation and Evolution,"
Nature Reviews Microbiology 3, no. 9 (September 2005): 679–87.

3. Rotem Sorek, Victor Kunin, and Philip Hugenholtz, "CRISPR—a Widespread
System That Provides Acquired Resistance Against Phages in Bacteria and
Archaea," *Nature Reviews Microbiology* 6, no. 3 (March 2008): 181–86.

4. Ward, *Life as We Do Not Know It*.

5. Centers for Disease Control and Prevention "Parasites—*Giardia*," cdc.gov/para
sites/giardia/general-info.html.

6. "What Does It Take to Kill a Waterbear (Tardigrade)?" *Quora*, www.quora.com
/What-does-it-take-to-kill-a-waterbear-tardigrade.

7. Paul Davies, *The Fifth Miracle: The Search for the Origin and Meaning of Life*
(New York: Simon & Schuster, 1999).

8. Ward, *Life as We Do Not Know It*.

9.   Jason Daley, "Behold LUCA, the Last Universal Common Ancestor of Life on Earth," *SmartNews* (blog), *Smithsonian.com*, July 26, 2017, smithsonianmag .com/smart-news/behold-luca-last-universal-common-ancestor-life-earth-180 959915.

10. Michael Le Page, "Universal Ancestor of All Life on Earth Was Only Half Alive," *New Scientist*, July 25, 2016, newscientist.com/article/2098564-universal -ancestor-of-all-life-on-earth-was-only-half-alive.

11. Charles Darwin to J. D. Hooker, February 1, 1871, Darwin Correspondence Project, https://www.darwinproject.ac.uk/letter/DCP-LETT-7471.xml.

12. The minimum number of genes reported here: Tina Hesman Saey, "Genes: How Few Needed for Life?" *Science News for Students*, April 5, 2016, www .sciencenewsforstudents.org/article/genes-how-few-needed-life.

13. Denyse O'Leary, "Life Continues to Ignore What Evolution Experts Say," *Evolution News*, September 9, 2015, evolutionnews.org/2015/09/life_forms_cont; Gogarten and Townsend, "Horizontal Gene Transfer, Genome Innovation and Evolution"; Csaba Pál, Balázs Papp, and Martin J. Lercher, "Adaptive Evolution of Bacterial Metabolic Networks by Horizontal Gene Transfer," *Nature Genetics* 37, no. 12 (December 2005): 1372–75.

14. Sir Archibald Geikie, "Lecture IV: Hutton's Fundamental Doctrines" and "Lecture VI: Lyell," *The Founders of Geology* (New York: Macmillan, 1897), 168, 281.

15. *Wikipedia*, s.v. "Circular Bacterial Chromosome," last modified December 18, 2017, en.wikipedia.org/wiki/Circular_bacterial_chromosome.

16. Eugene V. Koonin and Yuri I. Wolf, "Genomics of Bacteria and Archaea: The Emerging Dynamic View of the Prokaryotic World," *Nucleic Acids Research* 36, no. 21 (December 2008): 6688–719.

17. Anthony M. Poole, "Horizontal Gene Transfer and the Earliest Stages of the Evolution of Life," *Research in Microbiology* 160, no. 7 (September 2009): 473–80.

18. Jürgen Brosius, "Gene Duplication and Other Evolutionary Strategies: From the RNA World to the Future," *Journal of Structural and Functional Genomics* 3, nos. 1–4 (March 2003): 1–17.

19. Lynn Margulis, *Symbiosis in Cell Evolution: Life and Its Environment on the Early Earth* (Freeman: San Francisco, 1981); Maureen A. O'Malley, "Endosymbiosis and Its Implications for Evolutionary Theory," *Proceedings of the National Academy of Sciences of the United States of America* 112, no. 33 (August 2015), 10270–77.

20. Lynn Margulis, "Symbiogenesis and Symbionticism," *Symbiosis as a Source of Evolutionary Innovation: Speciation and Morphogenesis*, eds. Lynn Margulis and René Fester (Cambridge, MA: MIT Press, 1991), 1–13.

21. John Maynard Smith, "A Darwinian View of Symbiosis," *Symbiosis as a Source of Evolutionary Innovation: Speciation and Morphogenesis*, eds. Lynn Margulis and René Fester (Cambridge, MA: MIT Press, 1991), 26–39.

22. Nick Lane, *Life Ascending: The Ten Great Inventions of Evolution* (New York: W. W. Norton, 2009), 106.

23. "It Takes Teamwork: How Endosymbiosis Changed Life on Earth," Understanding Evolution, evolution.berkeley.edu/evolibrary/article/endosymbiosis_01.

CHAPTER VII: EPIGENETICS AND THE CAMBRIAN EXPLOSION

1. Douglas H. Erwin and James W. Valentine, *The Cambrian Explosion: The Construction of Animal Biodiversity* (Greenwood Village, CO: Roberts, 2013), 413.

2. On creationists and the Cambrian explosion, see "Does the Cambrian Explosion Pose a Challenge to Evolution?" *BioLogos*, biologos.org/common-questions/scientific-evidence/cambrian-explosion; David Campbell and Keith B. Miller, "The 'Cambrian Explosion': A Challenge to Evolutionary Theory?" in *Perspectives on an Evolving Creation*, ed. Keith B. Miller (Grand Rapids, MI: William B. Eerdmans, 2003), 182–204.

3. On Darwin and the base of Cambrian explosion, see Martin J. S. Rudwick, *The Meaning of Fossils: Episodes in the History of Paleontology* (New York: American Elsevier, 1972); Martin J. S. Rudwick, *Georges Cuvier, Fossil Bones, and Geological Catastrophes: New Translations and Interpretations of the Primary Texts* (Chicago: University of Chicago Press, 1998); James W. Valentine, *On the Origin of Phyla* (Chicago: University of Chicago Press, 2004).

4. Simon Conway Morris, *The Crucible of Creation: The Burgess Shale and the Rise of Animals* (Oxford, England: Oxford University Press, 1998).

5. On Charles Marshall and locomotion, see "Anima Interaction Behind 'Cambrian Explosion'?" *Harvard Gazette*, May 1, 2008, news.harvard.edu/gazette/story/2008/05/animal-interaction-behind-cambrian-explosion.

6. Simona Ginsburg and Eva Jablonka, "The Evolution of Associative Learning: A Factor in the Cambrian Explosion," *Journal of Theoretical Biology* 266, no 1 (September 2010): 11–20.

7. Andrew Parker, *In the Blink of an Eye: How Vision Sparked the Big Bang of Evolution* (New York: Basic Books, 2004)

8. Chris Phoenix, "Cellular Differentiation as a Candidate 'New Technology' for the Cambrian Explosion," *Journal of Evolution and Technology* 20, no. 2 (November 2009): 43–48.

9. Gáspár Jékely, Jordi Paps, and Claus Nielsen, "The Phylogenetic Position of Ctenophores and the Origin(s) of Nervous Systems," *EvoDevo* 6, no. 1 (January 2015): doi.org/10.1186/2041-9139-6-1.

10. Michael E. Baker, John W. Funder, and Stephanie R. Kattoula, "Evolution of Hormone Selectivity in Glucocorticoid and Mineralocorticoid Receptors," *Journal of Steroid Biochemistry and Molecular Biology* 137 (September 2013):

57–70; Joseph W. Thornton, "Evolution of Vertebrate Steroid Receptors from an Ancestral Estrogen Receptor by Ligand Exploitation and Serial Genome Expansions," *Proceedings of the National Academy of Sciences of the United States of America* 98, no. 10 (May 2001): 5671–76.

11. On unidirectional cell differentiation post-cnidarians, see Detlev Arendt et al., "The Evolution of Nervous System Centralization," *Philosophical Transactions of the Royal Society B: Biological Sciences* 363, no. 1496 (April 2008): 1523–28. See also the following review of stem cells and their workings: Sa Cai, Xiaobing Fu, and Zhiyong Sheng, "Dedifferentiation: A New Approach in Stem Cell Research," *BioScience* 57, no. 8 (September 2007): 655–62.

12. The venerable science publication *Nature* has bundled a number of important reviews on all aspects of the gut-brain axis and its workings at nature.com/collections/dyhbndhpzv.

13. R. M. Nesse, S. Bhatnagar, and E. A. Young, "Evolutionary Origins and Functions of the Stress Response," in *Encyclopedia of Stress*, 2nd ed., ed. George Fink (Waltham, MA: Academic Press, 2007), 965–70.

14. On cortisol's major role in fetal development of humans, see Elysia Poggi Davis and Curt A. Sandman, "The Timing of Prenatal Exposure to Maternal Cortisol and Psychosocial Stress Is Associated with Human Infant Cognitive Development," *Child Development* 81, no. 1 (January/February 2010): 131–48.

15. On stressors and epigenetics, see: Richard G. Hunter et al., R. Gagnidze, K. B. and D. Pfaff, "Stress and the Dynamic Dynamic Genome: Steroids, Epigenetics, and the Transposome," *Proceedings of the National Academy of Sciences of the United States of America* 112, no. 22 (June 2015): 6828–833.

CHAPTER VIII: EPIGENETICS PROCESSES BEFORE AND AFTER MASS EXTINCTIONS

1. J. John Sepkoski, "Mass Extinctions in the Phanerozoic Oceans: A Review," *Geological Society America*, Special Paper 190 (January 1982): 283–89; Peter D. Ward, *On Methuselah's Trail: Living Fossils and the Great Extinctions* (New York: W. H. Freeman, 1992). Further explored in David M. Raup, *The Nemesis Affair: A Story of the Death of Dinosaurs and the Ways of Science* (New York: W. W. Norton, 1986); Paul S. Martin and Richard G. Klein, eds., *Quaternary Extinctions: A Prehistoric Revolution* (Tucson: University of Arizona, 1984); Luis W. Alvarez et al., "Extraterrestrial Cause for the Cretaceous-Tertiary Extinction," *Science* 208, no. 4448 (June 1980): 1095–1108; Rick Gore, "Extinctions," *National Geographic*, June 1989: 663–99; David M. Raup, *Extinction: Bad Genes or Bad Luck?* (New York: W. W. Norton, 1991); Peter M. Sheehan

et al., "Sudden Extinction of the Dinosaurs: Latest Cretaceous, Upper Great Plains, U.S.A.," *Science* 254, no. 5033 (November 1991): 835–39; Jeff Hecht, "Asteroidal Bombardment Wiped Out the Dinosaurs," *New Scientist*, April 17, 1993, newscientist.com/article/mg13818692-200-science-asteroidal-bombard ment-wiped-out-the-dinosaurs.

2. On Siberian Traps and the Permian-Triassic mass extinction, see Becky Oskin, "Earth's Greatest Killer Finally Caught," *Live Science*, December 12, 2013, livescience.com/41909-new-clues-permian-mass-extinction.html; S. D. Burgess, J. D. Muirhead, and S. A. Bowring, "Initial Pulse of Siberian Traps Sills as the Trigger of the End-Permian Mass Extinction," *Nature Communications* 8 (July 2017): doi:10.1038/s41467-017-00083-9.

3. Peter Ward and Joe Kirschvink, *A New History of Life: The Radical New Discoveries About the Origins and Evolution of Life on Earth* (New York: Bloomsbury Press, 2015).

4. Peter Ward, *The End of Evolution: On Mass Extinctions and the Preservation of Biodiversity* (New York: Bantam, 1994); Edward O. Wilson, *The Diversity of Life* (Cambridge, MA: Harvard University Press, 1992).

5. Mother of All Mass Extinctions: Kristina Lapp, "The Mother of All Extinctions," *Actuality* (blog), August 22, 2010, actualityscience.blogspot.com/2010/07/mother -of-all-extinctions.html.

6. David M. Raup and J. John Sepkoski Jr., "Mass Extinctions in the Marine Fossil Record," *Science* 215, no. 4539 (March 1982): 1501–3.

7. Daniel H. Rothman et al., "Methanogenic Burst in the End-Permian Carbon Cycle," *Proceedings of the National Academy of Sciences of the United States of America* 111, no. 15 (April 2014): 5462–67.

8. Ibid.

9. The evolution of dogs is wonderfully summarized in a recent article in the *Atlantic*: Ed Yong, "A New Origin Story for Dogs," June 2, 2016.

10. Daniel Nätt et al., "Heritable Genome-Wide Variation of Gene Expression and Promoter Methylation Between Wild and Domesticated Chickens," *BMC Genomics* 13 (2012): doi: 10.1186/1471-2164-13-59.

11. Jinxiu Li et al., "Genome-Wide DNA Methylome Variation in Two Genetically Distinct Chicken Lines Using MethylC-seq," *BMC Genomics* 16 (2015): doi: 10 .1186/s12864-015-2098-8; Cencen Li et al., "Molecular Microevolution and Epigenetic Patterns of the Long Non-coding Gene *H19* Show Its Potential Function in Pig Domestication and Breed Divergence," *BMC Evolutionary Biology* 16 (2006): doi: 10.1186/s12862-016-0657-5.

12. Peter Ward and Alexis Rockman, *Future Evolution: An Illuminated History of Life to Come* (New York: W.H. Freeman, 2001).

13. Peter Ward, *Gorgon: Paleontology, Obsession, and the Greatest Catastrophe in Earth's History* (New York: Viking, 2004).

CHAPTER IX: THE BEST AND WORST OF TIMES
IN HUMAN HISTORY

1. Richard D. Alexander, *How Did Humans Evolve? Reflections on the Uniquely Unique Species* (Ann Arbor, MI: Museum of Zoology, University of Michigan); Mark V. Flinn, David C. Geary, and Carol V. Ward, "Ecological Dominance, Social Competition, and Coalitionary Arms Races: Why Humans Evolved Extraordinary Intelligence," *Evolution and Human Behavior* 26, no. 1: 10–46; Donald C. Johanson and Kate Wong, *Lucy's Legacy: The Quest for Human Origins* (New York: Three Rivers Press, 2010).

2. Carl Zimmer, "Siberian Fossils Were Neanderthals' Eastern Cousins, DNA Analysis Reveals," *New York Times*, December 22, 2010; David Reich et al., "Genetic History of an Archaic Hominin Group from Denisova Cave in Siberia," *Nature* 468, no. 1012 (December 2010): 1053–60.

3. M. H. Wolpoff et al., "Modern Human Origins," *Science* 241, no. 4867 (August 1998): 772–74; Ian Tattersall and Jeffrey H. Schwartz, "Hominids and Hybrids: The Place of Neanderthals in Human Evolution," *Proceedings of the National Academy of Sciences of the United States of America* 96, no. 13 (June 1999): 7117–19.

4. On the cognitive revolution, see: www.sciencedirect.com/topics/neuroscience /cognitive-revolution; Yuval Noah Harari, *Sapiens: A Brief History of Humankind* (New York: HarperCollins, 2015).

5. Ibid.; Erin Wayman, "Why Did the Human Mind Evolve to What It Is Today?" *Smithsonian.com*, June 25, 2012, smithsonianmag.com/science-nature/when-did -the-human-mind-evolve-to-what-it-is-today-140507905.

6. On the Toba eruption, see Marie D. Jones and John Savino, *Supervolcano: The Catastrophic Event That Changed the Course of Human History* (Franklin, NJ: Career Press, 2007), 140; C. A. Chesner et al., "Eruptive History of Earth's Largest Quaternary Caldera (Toba, Indonesia) Clarified," *Geology* 19, no. 3 (March 1991): 200–203.

7. Stanley H. Ambrose, "Late Pleistocene Human Population Bottlenecks, Volcanic Winter, and Differentiation of Modern Humans," *Journal of Human Evolution* 34, no. 6 (June 1998): 623–51; Michael R. Rampino and Stanley H. Ambrose, "Volcanic Winter in the Garden of Eden: The Toba Supereruption and the Late Pleistocene Human Population Crash," *Special Paper of the Geological Society of America* 345 (2000): doi: 10.1130/0-8137-2345-0.71; Michael R. Rampino and Stephen Self, "Climate-Volcanism Feedback and the Toba Eruption of ~74,000 Years Ago," *Quaternary Research* 40, no. 3 (November 1993): 269–80.

8. On CRII and the appearance of cave paintings, see Alexandra Kiely, "The Origin of the World's Prehistoric Cave Painting," *HeadStuff History* (blog), September 12, 2016, www.headstuff.org/history/origin-worlds-art-prehistoric-cave-painting; Harari, *Sapiens*.

9. Wayman, "Why Did the Human Mind Evolve to What It Is Today?"; Klein, R. G. & Edgar, B. (2002) *The Dawn of Human Culture* (Wiley, New York, 2002); Gregory Curtis, *The Cave Painters: Probing the Mysteries of the World's First Artists* (New York: Alfred A. Knopf, 2006).

10. Maina Kiarie, "Enkapune Ya Muto," *Enzi*, enzimuseum.org/peoples-cultures /your-genetic-history.

11. S. V. Zhenilo, A. S. Sokolov, and E. B. Prokhortchouk, "Epigenetics of Ancient DNA," *Acta Naturae* 8, no. 3 (July–September 2016): 72–76.

12. "What Does It Mean to Be Human?" *Smithsonian National Museum of Natural History*, humanorigins.si.edu/evidence/genetics.

13. David Gokhman et al., "Reconstructing the DNA Methylation Maps of the Neandertal and the Denisovan," *Science* 344, no. 6183 (May 2014): 523–27.

14. On the future human population maximum, see Brian Wang, "World Population Will Be Around 15–25 Billion in 2100 and Will Increase Through 2200 Because of African Fertility, Life Extension and Other Technology," *Nextbigfuture* (blog), August 27, 2015, nextbigfuture.com/2015/08/world-population-will -be-around-15-25.html.

15. Jim Erickson, "Washtenaw County Mammoth Find Hints at Role of Early Humans," *Michigan News* (blog), October 2, 2015, ns.umich.edu/new/multi media/videos/23181-washtenaw-county-mammoth-find-hints-at-role-of-early -humans; "Mammoth Article Q&A—Dr. Daniel Fisher, Renowned Paleontologist," *Mostly Mammoths, Mummies and Museums* (bog), September 10, 2013, mostlymammoths.wordpress.com/2013/09/10/mammoth-article-qa-dr-daniel -fisher-renowned-paleontologist.

16. Michael Sleza,, "Megafauna Extinction: DNA Evidence Pins Blame on Climate Change" *New* Scientist, July 23, 2015, newscientist.com/article/dn27952 -megafauna-extinction-dna-evidence-pins-blame-on-climate-change; Alan Cooper, Matthew Wooler, and Tim Rabanus Wallace, "How English-Style Drizzle Killed the Ice Age's Giants," *The Conversation*, April 18, 2017, thecon-versation.com/how-english-style-drizzle-killed-the-ice-ages-giants-76307.

17. On brain scans and neural plasticity, see Panagiotis Simos et al., "Insights into Brain Function and Neural Plasticity Using Magnetic Source Imaging," *Journal of Clinical* Neurophysiology 17, no. 2 (March 2000): 143–62; Federico Bermúdez-Rattoni, ed., *Neural Plasticity and Memory: From Genes to Brain Imaging* (Boca Raton, FL: CRC Press, 2007).

18. "Human History Timeline," *humanhistorytimeline.com.*

19. For more on the archeological timescale, see *Encyclopedia Britannica Online*, s.v. "Archaeogical Timescale," britannica.com/science/archaeological-timescale.

20. "The Big Five Personality Theory," *Positive Psychology Program*, June 23, 2017, positivepsychologyprogram.com/big-five-personality-theory.

21. On quantifying the Big Five personality traits, see Michael Gurven, "How Universal Is the Big Five? Testing the Five-Factor Model of Personality

Variation Among Forager-Farmers in the Bolivian Amazon," *Journal of Personality and Social Psychology* 104, no. 2 (February 2013): 354–70.

22.   On skeptics of personality entrained by genes, see Robert F. Krueger et al., "The Heritability of Personality Is Not Always 50%: Gene-Environment Interactions and Correlations Between Personality and Parenting," *Journal of Personality* 76, no. 6 (December 2008): 1485–1522.

23.   Zachary A. Kaminsky et al., "DNA Methylation Profiles in Monozygotic and Dizygotic Twins," *Nature Genetics* 41, no. 2 (February 2009): 240–45.

CHAPTER X: EPIGENETICS AND VIOLENCE

1.   See the U.S. Army War College's Strategic Studies Institute website: ssi .armywarcollege.edu.

2.   Tim Hetherington and Sebastian, dirs., *Restrepo* (Outpost Films, 2010). Sebastian Junger, "Into the Valley of Death," *Vanity Fair*, January 2008, discusses the strategic value of the Korangal Valley.

3.   Geerat J. Vermeij, "The Mesozoic Marine Revolution: Evidence from Snails, Predators and Grazers," *Paleobiology* 3, no. 3 (Summer 1977): 245–58; Steven M. Stanley, "Predation Defeats Competition on the Seafloor," *Paleobiology* 34, no. 1 (Winter 2008): 1–21.

4.   George P. Chrousos, "The Glucocorticoid Receptor Gene, Longevity, and the Complex Disorders of Western Societies," *American Journal of Medicine* 117, no. 3 (August 2004): 204–7.

5.   David M. Fergusson et al., "*MAOA*, Abuse Exposure and Antisocial Behaviour: 30-Year Longitudinal Study," *British Journal of Psychiatry* 198, no. 6 (May 2011): 457–63.

6.   Tony Merriman and Vicky A. Cameron, "Risk-taking: Behind the Warrior Gene Story," *New Zealand Medical Journal* 120, no. 1250 (March 2007): U2440; Kevin M. Beaver et al., "Exploring the Association Between the 2-Repeat Allele of the MAOA Gene Promoter Polymorphism and Psychopathic Personality Traits, Arrests, Incarceration, and Lifetime Antisocial Behavior," *Personality and Individual Differences* 54, no. 2 (January 2014): 164–68; Hayley M. Dorfman, Andreas Meyer-Lindenberg, and Joshua W. Buckholtz, "Neurobiological Mechanisms for Impulsive-Aggression: The Role of MAOA," in *Neuroscience of Aggression*, ed. Klaus A. Miczek and Andreas Meyer-Lindenberg (Berlin: Springer, 2013), 297–313, https://link.springer.com /chapter/10.1007%2F7854_2013_272#citeas.

7.   Rosie Mestel, "Does the 'Aggressive Gene' Lurk in a Dutch Family?" *New Scientist*, October 30, 1993, newscientist.com/article/mg14018970-600-does-the -aggressive-gene-lurk-in-a-dutch-family.

8.   Alondra Oubré, "The Extreme Warrior Gene: A Reality Check," *Scientia Salon* (blog), July 31, 2014, scientiasalon.wordpress.com/2014/07/31/the-extreme

-warrior-gene-a-reality-check/comment-page-1; Sarah Knapton, "Violence Genes May Be Responsible for One in 10 Serious Crimes," *Telegraph*, October 28, 2014, telegraph.co.uk/news/science/science-news/11192643/Violence-genes-may -be-responsible-for-one-in-10-serious-crimes.html.

9. "Licking Rat Pups: The Genetics of Nurture," *Nerve Blog*, November 11, 2010, sites.bu.edu/ombs/2010/11/11/licking-rat-pups-the-genetics-of-nurture.

10. "Baby's DNA Constructed Before Birth," *The Chart* (blog), CNN.com, June 7, 2012, thechart.blogs.cnn.com/2012/06/07/babys-dna-constructed -before-birth.

11. "Licking Rat Pups."

12. Aki Takahashi and Klaus A. Miczek, "Neurogenetics of Aggressive Behavior— Studies in Rodents" in Miczek and Meyer-Lindenberg, *Neuroscience of Aggression*, 3–44.

13. Child abuse statistics for USA: americanspcc.org/child-abuse-statistics/

14. Nils C. Gassen et al, "Life Stress, Glucocorticoid Signaling, and the Aging Epigenome: Implications for Aging-Related Diseases," *Neuroscience and Biobehavioral Reviews* 74B (March 2017): 356–65.

15. Anthony S. Zannas et al., "Lifetime Stress Accelerates Epigenetic Aging in an Urban, African American Cohort: Relevance of Glucocorticoid Signaling," *Genome Biology* 16 (December 2015): doi: 10.1186/s13059-015-0828-5.

16. Eric Turkheimer, "Three Laws of Behavior Genetics and What They Mean," *Current Directions in Psychological Science* 9, no. 5 (October 2000): 160–64.

17. Violence in the Middle Ages in described here: A. J. Finch, "The Nature of Violence in the Middle Ages: An Alternative Perspective," *Historical Research* 70, no. 173 (October 1997): 243–68.

18. Pieter Spierenburg, *Violence and Punishment* (Copenhagen: Polity Publishing, 2012).

19. Homicide rates New York City were profiled in *AM New York*. This is a big "sell" to the tourist trade: Anthony M. DeStefano, "New York City Homicide Rate Lowest Since Second World War," January 1, 2018.

20. The writings of Elias can be found here: www.norberteliasfoundation.nl

21. Homicide rates in Europe: https://www.indexmundi.com/facts/european-union /homicide-rate

22. Marco P. Boks et al., "Longitudinal Changes of Telomere Length and Epigenetic Age Related to Traumatic Stress and Post-traumatic Stress Disorder," *Psychoneuroendocrinology* 51 (January 2015): 506–12; Amy L. Non et al., "DNA Methylation at Stress-Related Genes Is Associated with Exposure to Early Life Institutionalization," *American Journal of Physical Anthropology* 161, no. 1 (September 2016): 84–93.

23. FBI crime statistics can be found here: www.ucrdatatool.gov. But see also a review of how the administration of President Trump plays with the FBI crime statistics: Clare Malone and Jeff Asher, "The First FBI Crime Report Issued

Under Trump Is Missing a Ton Of Info," October 27, 2017, *FiveThirtyEight*, fivethirtyeight.com/features/the-first-fbi-crime-report-issued-under-trump-is -missing-a-ton-of-info/

24. Number of U.S. soldiers who could not kill: www.historynet.com/men-against -fire-how-many-soldiers-actually-fired-their-weapons-at-the-enemy-during-the -vietnam-war.htm.

CHAPTER XI: CAN FAMINE AND FOOD CHANGE OUR DNA?

1. Nessa Carey, *The Epigenetics Revolution: How Modern Biology Is Rewriting Our Understanding of Genetics, Disease, and Inheritance* (New York: Columbia University Press, 2012).

2. A detailed scientific article on the Dutch Hunger can be found in Laura C. Schultz, "The Dutch Hunger Winter and the Developmental Origins of Health and Disease," *Proceedings of the National Academy of Sciences of the United States of America* 107, no. 39: 16757–58.

3. Aryeh D. Stein and L. H. Lumey, "The Relationship Between Maternal and Offspring Birth Weights After Maternal Prenatal Famine Exposure: The Dutch Famine Birth Cohort Study," *Human Biology* 72, no. 4 (August 2000): 641–54.

4. Nicky Hart, "Famine, Maternal Nutrition and Infant Mortality: A Re-examination of the Dutch Hunger Winter," *Population Studies* 47, no. 1 (March 1993): 27–46.

5. Alan S. Brown and Ezra S. Susser "Prenatal Nutritional Deficiency and Risk of Adult Schizophrenia," *Schizophrenia Bulletin* 34, no. 6 (November 2008): 1054–63.

6. Oded Rechavi et al., "Starvation-Induced Transgenerational Inheritance of Small RNAs in *C. elegans*," *Cell* 158, no. 2 (July 2017): 277–87.

7. Adelheid Soubry et al., "Obesity-Related DNA Methylation at Imprinted Genes in Human Sperm: Results from the TIEGER Study," *Clinical Epigenetics* 8 (May 2016): doi: 10.1186/s13148-016-0217-2.

8. David Epstein, "How an 1836 Famine Altered the Genes of Children Born Decades Later," *io9*, August 26, 2013, io9.gizmodo.com/how-an-1836-famine -altered-the-genes-of-children-born-d-1200001177.

9. G. Kaati, L. O. Bygren, and S. Edvinsson, "Cardiovascular and Diabetes Mortality Determined by Nutrition During Parents' and Grandparents' Slow Growth Period," *European Journal of Human Genetics* 10, no. 11 (November 2002): 682–88.

10. "Famine in China," in *Encyclopedia of Population*, ed. Paul Demeny and Geoffrey McNicoll, Geoffrey (New York: Macmillan Reference, 2003), 1:388–90.

11. Leslie Holmes, *Communism: A Very Short Introduction* (Oxford University Press 2009), 32.

12. Benjamin Reilly, *Disaster and Human History: Case Studies in Nature, Society and Catastrophe* (Jefferson, NC: McFarland, 2009), 281.

13. Brian K. Hall, *Evolutionary Developmental Biology*, 2nd ed. (Dordrecht, Netherlands, 1999), 328.

14. Andrew B. Shreiner, John Y. Kao, and Vincent B. Young, "The Gut Microbiome in Health and in Disease," *Current Opinion in Gastroenterology* 31, no. 11 (January 2015): 69–75.

15. David R. Montgomery and Anne Biklé, *The Hidden Half of Nature: The Microbial Roots of Life and Health* (New York: W. W. Norton, 2015).

16. Mark Roth, "Suspended Animation Is Within Our Grasp," Ted Talk, February 2010, www.ted.com/talks/mark_roth_suspended_animation.

17. Richard G. Hunter, "Epigenetic Effects of Stress and Corticosteroids in the Brain," *Frontiers in Cellular Neuroscience* 6 (April 2012): doi: 10.3389/fncel.2012 .00018.

CHAPTER XII: THE HERITABLE LEGACY
OF PANDEMIC DISEASES

1. On bleeding of patients, see Jennie Cohen, "A Brief History of Bloodletting," *History Stories* (blog), History.com, May 30, 2012, history.com/news/a-brief -history-of-bloodletting.

2. A good introduction to human epidemics through time is here: "Deadly Diseases: Epidemics throughout History," CNN, www.cnn.com/interactive /2014/10/health/epidemics-through-history/. A visual history is here: Charlie Sorrel, "The Epidemics That Have Defined Human History, in One Chart," *Fast Company*, March 2, 2016, www.fastcompany.com/3057256/the-epidemics -that-have-defined-human-history-in-one-chart

3. The role of iron in combatting disease, especially the Bubonic Plague, is well described here: Bradley Wertheim, "The Iron in Our Blood That Keeps and Kills Us," *Atlantic*, January 10, 2013, https://www.theatlantic.com/health/ archive/2013/01/the-iron-in-our-blood-that-keeps-and-kills-us/266936/

4. The specific times and places of the Bubonic Plague are listed here: www .history.com/topics/black-death

5. The debilitating effects of stress are discussed in a previous chapter. One of the best sources is George P. Chrousos, "The Glucocorticoid Receptor Gene, Longevity, and the Complex Disorders of Western Societies," *American Journal of Medicine* 117, no. 3 (August 2004): 204–7.

6. On the God gene and VMAT, see Linda A. Silveira, "Experimenting with Spirituality: Analyzing *The God Gene* in a Nonmajors Laboratory Course," *CBE-Life Sciences Education* 7, no. 1 (Spring 2008): 132–45; Dean Hamer, *The God Gene: How Faith Is Hardwired into Our Genes* (New York: Anchor, 2005); P. Z. Myers, "No God, and No 'God Gene,' Either," *Pharyngula*, February 13, 2005, archived from the original on October 3, 2009, http://web.archive.org/web/20091003 213607/http://pharyngula.org/index/weblog/comments/no_god_and_no_god

_gene_either/; Carl Zimmer, "Faith-Boosting Genes: A Search for the Genetic Basis of Spirituality," *Scientific American*, October 2004.

7.   "Neurotheology: This Is Your Brain on Religion," *Talk of the Nation*, December 15, 2010, npr.org/2010/12/15/132078267/neurotheology-where-religion-and -science-collide.

8.   Ibid.; the observed effects of the brains of self-described atheists: blog.al.com /wire/2014/01/religious_brains_function_diff.html

9.   On VMAT and production of hormones, see "University Hospital, Marseille: Expression of Somatostatin Receptors, Dopamine D2 Receptors, Noradrenaline Transporters, and Vesicular Monoamine Transporters in 52 Pheochromocytomas and Paragangliomas," in *Pituitary Hormone Release Inhibiting Hormones—Advances in Research and Application*, Q Ashton Acton, ed. (Atlanta: ScholarlyEditions, 2012), 62–63.

10. Leah Marieann Klett, "Groundbreaking New Study Finds Cancer Patients Who Believe in God Experience Less Symptoms, Greater Emotional Health," *Gospel Herald*, August 11, 2015, gospelherald.com/articles/57069/20150811/ground breaking-new-study-finds-cancer-patients-who-believe-in-god-experience-less -symptoms-greater-emotional-health.htm.

CHAPTER XIII: THE CHEMICAL PRESENT

1.   For more on the history of the study of epigenetics from toxins, see Andrea Baccarelli and Valentina Bollati, "Epigenetics and Environmental Chemicals," *Current Opinion in Pediatrics* 21, no. 2 (April 2009): 243–51.

2.   On the increase in the level of chemicals in environment over the past few decades and the effects, see Stella Marie Reamon-Buettner, Vanessa Mutschler, and Juergen Borlak, "The Next Innovation Cycle in Toxicogenomics: Environmental Epigenetics," *Mutation Research* 659, no. 1–2 (July–August 2008): 158–65; Randy L. Jirtle and Michael K. Skinner, "Environmental Epigenomics and Disease Susceptibility," *Nature Reviews: Genetics* 8, no. 4 (April 2007): 253–62.

3.   Bruce A. Fowler et al., "Oxidative Stress Induced by Lead, Cadmium and Arsenic Mixtures: 30-Day, 90-Day, and 180-Day Drinking Water Studies in Rats: An Overview," *Biometals* 17, no. 5 (October 2004): 567–68.

4.   Samuel M. Goldman, "Environmental Toxins and Parkinson's Disease," *Annual Review of Pharmacology and Toxicology* 54 (2014): 141–64.

5.   Pan Chen et al., "Age- and Manganese-Dependent Modulation of Dopaminergic Phenotypes in *C. Elegans* DJ-1 Genetic Model of Parkinson's Disease," *Metallomics* 7, no. 2 (February 2015): 289–98.

6.   Chemicals found in human mothers' milk can be found in a recent report in *Forbes* magazine: Tara Haelle, "How Toxic Is Your Breastmilk?" August 21, 2015, www.forbes.com/sites/tarahaelle/2015/08/21/how-toxic-is-your-breastmilk/

7. Sperm counts in American males are dropping: Rob Stein, "Sperm Counts Plummet in Western Men, Study Finds," NPR, July 31, 2017, www.npr.org/2017 /07/31/539517210/sperm-counts-plummet-in-western-men-study-finds.

8. The top producers of pesticides in 2015: California Department of Pesticide Regulation, "Pesticide Use Reporting—2015 Summary Data," www.cdpr.ca.gov /docs/pur/pur15rep/15_pur.htm.

9. Michael K. Skinner, "A New Kind of Inheritance," *Scientific American*, August 2013, 44–51. Randy L. Jirtle and Michael K. Skinner, "Environmental Epigenomics and Disease Susceptibility," *Nature Reviews: Genetics* 8, no. 4 (April 2007): 253–62. An article from Skinner's university profiling his work can be found here: Eric Sorensen, "WSU Researchers See Popular Herbicide Affecting Health Across Generations," *WSU News*, September 20, 2017, news .wsu.edu/2017/09/20/herbicide-affecting-health-across-generations; his complete list of publications here: skinner.wsu.edu/publications; there is also a video produced by *Scientific American* included in this article on Skinner's work: W. Wayt Gibbs, "Can We Inherit the Environmental Damage Done to Our Ances-tors," ScientificAmerican.com, July 15, 2014, scientificamerican.com/article/can -we-inherit-the-environmental-damage-done-to-our-ancestors-video. But much more entertaining, in that it contains good Skinner quotes, is a piece that appeared in *Smithsonian*: Jeneen Interlandi, "The Toxins That Affected Your Great-grandparents Could Be in Your Genes," *Smithsonian*, December 2013, smithsonianmag.com/innovation/the-toxins-that-affected-your-great -grandparents-could-be-in-your-genes-180947644.10. Daniel Beck, Ingrid Sadler-Riggleman, and Michael K. Skinner, "Generational Comparisons (F1 Versus F3) of Vinclozolin Induced Epigenetic Transgenerational Inheritance of Sperm Differential DNA Methylation Regions (Epimutations) Using MeDIP-Seq," *Environmental Epigenetics* 3, no. 3 (July 2017): 1–12.

10. Current pot use in America: www.cdc.gov/mmwr/volumes/65/ss/ss6511a1.htm, and Christopher Ingraham, "11 Charts That Show Marijuana Has Truly Gone Mainstream," *Washington Post*, April 19, 2017, www.washingtonpost.com/news /wonk/wp/2017/04/19/11-charts-that-show-marijuana-has-truly-gone -mainstream/?utm_term=.b66a4ff54a0b

11. Henrietta Szutorisz and Yasmin L. Hurd, "Epigenetic Effects of Cannabis Expo-sure," *Biological Psychiatry* 79, no. 7 (April 2016): 586–94.

12. Yasmin L. Hurd, "Multigenerational Epigenetic Effects of Cannabis Exposure," *Grantome*, National Institutes of Health (2012–17), http://grantome.com/grant /NIH/R01-DA033660-01.

13. On nicotine and fertility, see Bailey Kirkpatrick, "Nicotine Could Cause Epigenetic Changes to Testes and Compromise Fertility," *What Is Epigenetics* (blog), March 29, 2016, whatisepigenetics.com/nicotine-could-cause-epigene tic-changes-to-testes-and-compromise-fertility; on smoking, epigenetics, and cancer: Chien-Hung Lee et al., "Independent and Combined Effects of Alcohol

Intake, Tobacco Smoking and Betel Quid Chewing on the Risk of Esophageal Cancer in Taiwan," *International Journal of Cancer* 113, no. 3 (January 2005): 475–82; Yiping Huang et al., "Cigarette Smoke Induces Promoter Methylation of Single-Stranded DNA-Binding Protein 2 in Human Esophageal Squamous Cell Carcinoma," *International Journal of Cancer* 128, no. 10 (May 15, 2011): 2261–73.

14. On epigenetics and hard drug use, see David A. Nielsen et al., "Epigenetics of Drug Abuse: Predisposition or Response," *Pharmacogenomics* 13, no. 10 (August 2012): 1149–60.

15. James P. Curley, Rahia Mashoodh, and Frances A. Champagne, "Epigenetics and the Origins of Paternal Effects," *Hormones and Behavior* 59, no. 3 (March 2011): 306–14.

16. Ibid. The effect of betel nut cannot be overestimated: www.webmd.com/vitamins/ai/ingredientmono-995/betel-nut.

17. Ernest Abel "Paternal Contribution to Fetal Alcohol Syndrome," *Addiction Biology* 9, no. 2 (June 2004): 127–33, 135–36.

18. Abigail Tracy, "This May Be the Lamest Viral-Marketing Campaign Ever," *Vocativ*, September 26, 2014, vocativ.com/money/business/gestations-bar-for -pregnant-women-viral-marketing-stunt/index.html.

CHAPTER XIV: FUTURE BIOTIC EVOLUTION IN THE CRISPR-CAS9 WORLD

1. Gregory Cochran and Henry Harpending, *The 10,000 Year Explosion: How Civilization Accelerated Human Evolution* (New York: Basic Books, 2010). On Harpending: https://www.splcenter.org/fighting-hate/extremist-files/individual /henry-harpending.

2. Luis B. Barreiro et al., "Natural Selection Has Driven Population Differentiation in Modern Humans," *Nature Genetics* 40, no. 3 (March 2008): 340–45.

3. Megan Gannon, "Race Is a Social Construct, Scientists Argue," *Scientific American*, February 5, 2016, scientificamerican.com/article/race-is-a-social-const ruct-scientists-argue.

4. Jeffrey C. Long and Rick A. Kittles, "Human Genetic Diversity and the Nonexistence of Biological Races," *Human Biology* 75, no. 4 (August 2003): 449–71.

5. Elizabeth Weise, "Sixty Percent of Adults Can't Digest Milk," *USA Today*, abcnews.go.com/Health/WellnessNews/story?id=8450036.

6. On Lassa fever resistance from natural selection: Kristian G. Andersen, "Genome-Wide Scans Provide Evidence for Positive Selection of Genes Implicated in Lassa Fever," *Philosophical Transactions of the Royal Society B: Biological Sciences* 367, no. 1590 (March 2012): 868–77.

7. On the danger of CRISPR, see Heidi Ledford, "CRISPR, the Disruptor," *Nature* 522, no. 7554 (June 2015): 20–24. As a contrast, see this 2017 propaganda piece from a chemical engineering trade publication: Melody M. Bomgardner,

"CRISPR: A New Toolbox for Better Crops," *Chemical & Engineering News* 95, no. 24 (June 2017): 30–34.

8. For a threat assessment on CRISPR as a means of producing bioweapons, see Antonio Regalado, "Top U.S. Intelligence Official Calls Gene Editing a WMD Threat," *MIT Technology Review*, February 9, 2016, technologyreview.com/s /600774/top-us-intelligence-official-calls-gene-editing-a-wmd-threat.

9. George Dvorsky, "Gene-Edited Dogs with Jacked-Up Muscles Are a World's First," *Gizmodo*, October 20, 2015, gizmodo.com/gene-edited-dogs-with-jacked -up-muscles-are-a-worlds-fi-1737545538.

10. For photographs of the Chinese Frankenbeagles, see Tina Heisman Saey, "Muscle-Gene Edit Creates Buff Beagles," *Science News*, October 23, 2015, sciencenews.org/article/muscle-gene-edit-creates-buff-beagles.

11. The $100 million U.S. military investment (so far) in designer creatures for warfare.

12. Paul A. Philips, "DARPA: Genetically Modified Humans for a Super Soldier Army," *Activist Post*, October 11, 2015, activistpost.com/2015/10/darpa-gene tically-modified-humans-for-a-super-soldier-army.html.

13. Shivali Best, "Genetically-Modified Superhuman Soldiers of the Future Will Feel No Pain or Fear and Be More 'Destructive' Than Nuclear Bombs,' Warns Vladimir Putin," *Daily Mail*, October 23, 2017, dailymail.co.uk/sciencetech /article-5008461/Vladimir-Putin-warns-super-human-soldiers-future.html #ixzz52nb4njna.

14. Kelly Servick, "First U.S. Team to Gene-Edit Human Embryos" *Science*, July 27, 2017, sciencemag.org/news/2017/07/first-us-team-gene-edit-human-embryos -revealed.

EPILOGUE: LOOKING FORWARD

1. American sperm counts dropping over last 40 years: Rob Stein, "Sperm Counts Plummet in Western Men, Study Finds," NPR, July 31, 2017, www.npr.org/2017 /07/31/539517210/sperm-counts-plummet-in-western-men-study-finds.

2. The profile of Mike Skinner: Eric Sorensen, "WSU Researchers See Popular Herbicide Affecting Health Across Generations," *WSU News*, September 20, 2017,

3. The Intergovernmental Panel on Climate Change (IPCC): www.ipcc.ch

4. Peter Ward, *Our Flooded Earth* (New York: Basic Books, 2012).

5. The new maps from 2017 model those places on Earth that would experience more than two hundred days per year of above 100-degree-Fahrenheit temper-atures, or the nearly equivalent, 40-degree-Celsius mark. The model was a predic-tion for the year 2100. Heat is coming, as is moisture in the form of monstrous and human-killing rainstorm events and seasons that are still limited today to the areas that have long experienced a climate called "monsoon." Endless rainfall

for half the year. Amid killing heat. For the U.S.: www.climatecentral.org/news
/summer-temperatures-co2-emissions-1001-cities-16583. For a global picture:
mashable.com/2017/07/06/climate-change-shifting-cities-hotter-summers
/#INHvb2sxEiq3.

6. The extraordinary flooding of Houston in 2017: Jason Samenow, Angela Fritz,
and Greg Porter, "Catastrophic Flooding 'Beyond Anything Experienced' in
Houston and 'Expected to Worsen,'" *Washington Post*, August 27, 2017, www
.washingtonpost.com/news/capital-weather-gang/wp/2017/08/27/catastrophic
-flooding-underway-in-houston-as-harvey-lingers-over-texas/.

7. One of the most important yet overlooked scientific discoveries is that stress
increases mutation rate, and that increases evolutionary rate. Things in a
stressful environment evolve faster: Rodrigo Galhardo, P. Hastings and Susan
Rosenberg. "Mutation as a Stress Response and the Regulation of Evolution,"
Critical Reviews Biochemistry and Molecular Biology 42 (2007): 399–435.

8. Thomas Malthus, *An Essay on the Principle of Population . . .* (London:
J. Johnson, 1798).

9. Daniel Kolitz, "Can Superhuman Mutants Be Living Among Us?" *Gizmodo*,
June 5, 2017, gizmodo.com/can-superhuman-mutants-be-living-amongst-us
-1795696308.

10. On the bad professor who suggested America is experiencing a stress epidemic,
see Dan Jackson, "Prof: 'Stress' of Trump Election Will Change Human Evolution," *Campus Reform*, June 8, 2017, https://www.campusreform.org/?ID=9284;
Nicholas Staropoli, "Epigenetics Around the Web: Trump Isn't Affecting Human
Evolution and Organic Produce Isn't Helping Your Sperm," *Epigenetics Literacy
Project*, June 19 2017, epigeneticsliteracyproject.org/blog/epigenetics-around
-web-trump-isnt-affecting-human-evolution-organic-produce-isnt-helping
-sperm.

11. Michael Sleza, "Megafauna Extinction: DNA Evidence Pins Blame on Climate
Change," *New Scientist*, July 23, 2015, newscientist.com/article/dn27952
-megafauna-extinction-dna-evidence-pins-blame-on-climate-change; Alan
Cooper, Matthew Wooler, and Tim Rabanus Wallace, "How English-Style
Drizzle Killed the Ice Age's Giants," *The Conversation*, April 18, 2017, thecon
versation.com/how-english-style-drizzle-killed-the-ice-ages-giants-76307.

12. Arthur C. Clarke, "Hazards of Prophecy: The Failure of Imagination," *Profiles
of the Future: An Inquiry into the Limits of the Possible* (New York: Harper &
Row, 1962), 14.

Index

acetate and acetate-processing genes,
    130–31, 132, 187
acritarchs, 90
ACTH (adrenocorticotropic hormone), 121
adaptation
    evolution as key to, 98
    generation of traits and, 49–50, 82–83
    herbivores and carnivore communities
        and, 140–41
    human races and, 211
    natural selection and, 48, 50, 98
    river morphology changes and, 139
ADD (attention deficit disorder), 208
Adenosine triphosphate (ATP), 122
adrenal glands, 121
adrenaline, 117, 167, 194
adrenocorticotropic hormone (ACTH), 121
Advanced Tools for Mammalian Genome
    Engineering, 218
agreeableness (personality trait), 160–61
agriculture. *See also* food and famine
    farmers, observations of, 13–14, 16–17,
        34–35
    first agricultural states, 165
    human evolution and, 137, 143–44,
        155–56, 164–65
    invention of, 153, 155
    plagues and, 193
    sea levels rising and, 36
    toxins, agricultural, 204–6
AI (artificial intelligence), 1, 97
air pollution, 200, 201, 204
albedo change, 148
alcoholism. *See* drug and alcohol abuse
allopatric speciation, 45, 47
altruism, 116, 117
Alvarez, Luis, 45–46, 124
Alvarez, Walter, 124
ammomites, 14, 28
amoebae, 109
amphioxus, 119
anemones, 117, 118
animals
    chemicals found in, 204
    domestication of, 34–36, 43, 132–35, 137,
        142–43, 160
    extinction susceptibility, 138

first appearance of, 91
gene imprinting in, 106
genetic engineering of, 215, 216–18
gut biomes of, 156, 185–87
phyla, and Cambrian explosion, 113,
    117, 123
transition to life on land, 80–81, 91
Anomalocaris, 120
Anthropocene epoch, 136–37
antioxidants, 188–89
Antonine Plague (A.D. 165–80), 192
archaea, 76, 101, 106–7
arsenic, 200, 204
art, sudden appearance of, 150, 151
arthropods, 114, 119, 120, 165
artificial intelligence (AI), 1, 97
Asaro, Frank, 124
asteroid impacts, 18–19, 76–77, 81–82,
    87–88, 93, 124, 127–29, 135, 151
atheists, 196
atomic bomb, 221–22, 224
ATP (Adenosine triphosphate), 122
attention deficit disorder (ADD), 208
Australopithecus afarensis, 145

baby boomers, 178
Baccarelli, Andrea "Epigenetics and
    Environmental Chemicals," 200–202
bacteria. *See also* microbes
    amoebae invaded by, 109–10
    diseases and death caused by, 70–71, 190,
        191–92
    diversification of, 104–6
    DNA of, 102, 105, 213–14
    global diversity and, 128–29
    in gut biomes, 186
    heritable and non-heritable DNA states of,
        106
    hydrogen sulfide and, 189
    lateral gene transfer and, 95, 101, 102, 103
    photosynthetic, 89, 104, 105
    RNAi and, 76
    viruses and, 75
behavior
    changes in human, 143–44, 146, 147, 151,
        156, 160–62, 194
    chemical exposure and, 201, 207

A Note on the Author

Peter Ward, PhD, is a paleontologist and astrobiologist. His previous book is *A New History of Life*, with co-author Joe Kirschvink; Ward's *Rare Earth*, co-authored with Donald Brownlee, was named by *Discover* magazine as one of the ten most important science books of 2001, and his *Gorgon* was awarded a Washington State Book Award in 2005. He has appeared often on the Art Bell *Coast to Coast AM* radio program and on *Science Friday* with Ira Flatow. Ward lives in Washington State.